OUR SUN

⊙   ⊙

⊙

THE HARVARD BOOKS ON ASTRONOMY

Edited by HARLOW SHAPLEY *and* CECILIA PAYNE-GAPOSCHKIN

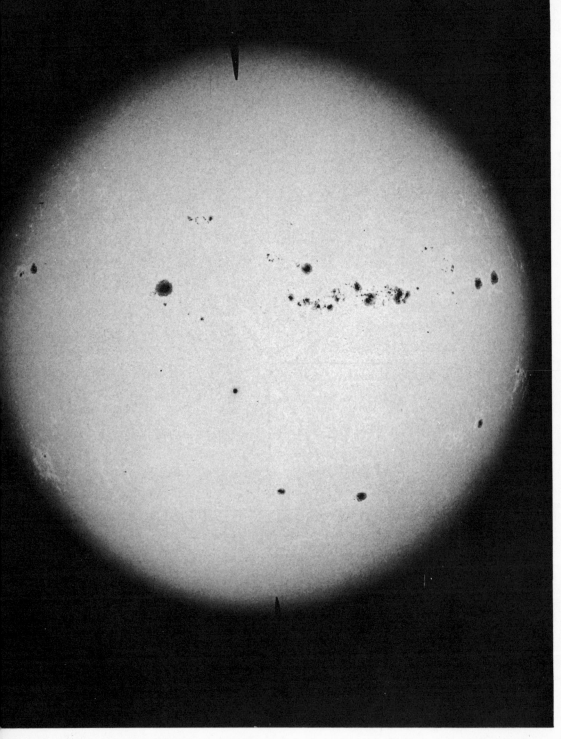

Heavily spotted sun, 14 August 1947. The dark shadows represent the north and south points of the solar disk. (Mount Wilson Observatory.)

Donald H. Menzel

OUR SUN

⊙ ⊙

REVISED EDITION

⊙

HARVARD UNIVERSITY PRESS

Cambridge, Massachusetts

1959

ASTRONOMY

© Copyright 1949, 1959 by the President
and Fellows of Harvard College

Distributed in Great Britain by

Oxford University Press   ·   London

Library of Congress Catalog Card Number 59–12975
Printed in the United States of America

A note about the manufacture and design of this book: No conventional types or engravings were used in the making of this book. All type matter was photocomposed. Type on film (combined with photographic negatives of the illustrations) was used in making photolithographic printing plates. The types used are Intertype Fotosetter Baskerville, Futura Medium, and Futura Demibold. The phototypesetting and film stripping were performed by Graphic Services, Inc., York, Pennsylvania.

The plates and printing are the work of The Murray Printing Company, Forge Village, Massachusetts. The paper is a special grade of offset stock manufactured by the P. H. Glatfelter Company of Spring Grove, Pennsylvania. A book cloth fabric made by the Holliston Mills, Norwood, Massachusetts, was used by the Stanhope Bindery, of Boston, Massachusetts, in the binding.

The book was designed by and produced under the direction of Burton L. Stratton with the assistance of Marcia R. Lambrecht, members of the Harvard University Press production staff.

Dedication

In recognition of their significant contributions to problems of solar astronomy, I dedicated the original edition of this book to my good friends and colleagues, the following astronomers and astrophysicists of France:

Lucien d'Azambuja, *Observatoire de Meudon*
Marguerite d'Azambuja, *Observatoire de Meudon*
Daniel Barbier, *Institute d'Astrophysique*
Daniel Chalonge, *Institute d'Astrophysique*
Bernard Lyot, *Observatoires Meudon, Pic du Midi.*

The tragic death of Bernard Lyot, while en route home after observing the solar eclipse of 25 February 1952 in Africa, was a great loss to solar astronomy.

Preface

The ten years that have elapsed since I published the first edition of *Our Sun* have been fruitful in terms of solar research. Spectrographs and telescopes sent aloft in rockets have contributed important facts. Observations from new and more powerful coronagraphs and advances in theoretical interpretation of the data have altered many of our concepts of the sun and solar activity. For example, we can no longer subscribe to the hypothesis that sunspots are storms, resembling terrestrial cyclones. Instead, we find that spots are the most quiet regions of the entire solar surface, stabilized by the presence of strong magnetic fields. The most unstable regions of the solar atmosphere occur near the spots' periphery. Here, rising bubbles of hot gas acquire enormous energies and heat the solar corona to temperatures in excess of one million degrees Centigrade.

The U.S. Air Force has supported basic solar researches, not only at Sacramento Peak Observatory, but also at the Harvard Radio Astronomy Station at Fort Davis, Texas, and at Harvard College Observatory in Cambridge, Massachusetts.

For illustrative matter and for basic data I am particularly indebted to my colleagues of the Sacramento Peak Observatory, especially Dr. John W. Evans, Dr. Frank Q. Orrall, Dr. Henry J. Smith, Richard B. Dunn, Harry Ramsey, and Howard de Mastus; and to Dr. Alan Maxwell at Fort Davis. Dr. Richard Tousey, of the U.S. Naval Research Laboratory, has generously supplied information in advance of publication.

I am grateful for the continued support of the various institutions that provided data and pictures for the first edition. I regret that lack of space forces me to curtail references to specific individuals.

The institutions that have contributed include: Sacramento Peak Observatory, Fort Davis Radio Astronomy Station, High Altitude Observatory of the University of Colorado, Observatoires de Meudon and Pic du Midi, Institute d'Astrophysique, McMath-Hulbert Observatory of the University of Michigan, Mount Wilson Observatory, Lick Observatory, Sproul Observatory, Yerkes Observatory, Lowell Observatory, U.S. Naval Observatory, Cornell University, U.S. Naval Research Laboratory, National Research Council of Canada, Radiophysics Laboratory of the Commonwealth Scientific and Industrial Research Organization of Australia, Smithsonian Institution, Carnegie Institution of Washington, the Spectroscopic Laboratory and the Solar Energy Project of Massachusetts Institute of Technology, National Bureau of Standards, National Geographic Society, U.S. Navy, U.S. Army, Harvard Physics Laboratory, Princeton University Observatory, General Mills Corporation, Prentice-Hall, Henry Holt & Co., *Sky and Telescope*, Baird Atomic, Radio Corporation of America, Avco Research Laboratory, University of California, Brookhaven National Laboratory, California Institute of Technology.

At Harvard College Observatory, Dr. Thomas Gold, Dr. Max Krook, and Dr. David Layzer generously provided information. Above all, I wish to thank Dr. Barbara Bell who has assisted me efficiently in every phase of this new edition, from the collecting and sorting of facts to the reading of the final proof.

D. H. M.

Contents

Our Sun

1

Meet the Sun

The sun should require no introduction. Daily it rises and daily it sets. In the spring it swings northward in the sky and in the autumn it retires toward the south. The position of the sun can be predicted with great accuracy many thousands of years in the future. Because of this regularity in motion, we have become all too accustomed to take the sun for granted.

Frozen in the rocks is a record of another sort of solar dependability: fossils showing the unbroken history of life from almost the first delicate protozoan cell through the various evolutionary forms down (or rather up) to our modern era. The dating of rock formations is not an easy matter, resting as it does upon estimates of the rates of deposition of silt or upon determinations of the quantity of helium and radiolead, the ashes of radioactive atoms that "burn" themselves up at a known, unvarying rate. New techniques, which permit the detection of minute traces of atomic species, have, however, led to increasing refinements of our measures.

The best evaluations place the origin of life at more than 500,000,000 years ago. For at least this period of time, and probably much longer, the sun has been radiating from its seemingly inexhaustible store of light and heat. And we may be sure that through all those ages the sun has never so much as halved or doubled its output of energy. Otherwise, the fossil record would show abrupt discontinuities.

But have we the right to assume that the sun will continue to shine forever? Can we even be certain that tomorrow the sun will not fade away to a fraction of its accustomed brilliance, or perhaps explode with a violence so great that the entire earth will be wrecked and vaporized in the outburst? For the sun is just a star—one of the hundred thousand million stars that comprise our great Milky Way. And among those stars we do find occasional examples that behave catastrophically—stars that fade at unpredictable intervals and stars that explode. Are we sure that our sun is immune from either of these peculiar stellar afflictions? To find the answer—if we can—we must look to the sun itself.

The sun, like a dependable and steady husband, is rarely fully appreciated. Its very regularity precludes our taking very much notice of it. Yet, think for a moment what this world would be like if the sun were suddenly to vanish from the sky! The earth would be plunged into immediate darkness, lit dimly by only the stars. The moon and planets would likewise be extinguished, for they shine only by reflected sunlight. Arctic chill would grip the entire earth. Within a week the tropics would be snowbound. The rivers would cease to flow. The winds would stop and gradually the oceans would freeze to their depths.

Man, bereft of direct solar heat, would suddenly discover, if he had not previously realized it, that water, wind, and tide power are absolutely dependent on the sun for existence. And vegetation, another source of energy, would cease to grow. Man could not long continue to exist against such overwhelming odds. Here and there, with the aid of petroleum, coal, or wood—whose energy values also came originally from the sun—man might be able to postpone, but only slightly, his inevitable extinction. In time, the gases of the atmosphere would liquefy and freeze upon the lifeless world, as the temperature gradually sank to a few degrees above the absolute zero. An immense glacier of solid air, averaging 23 feet in thickness, would eventually encase the earth.

The cold catastrophe appears to be inescapable—sometime— unless the earth were to suffer a premature and more abrupt demise by collision with a sizable wandering planet or star, or by liquidation and evaporation if the sun chanced to explode and become a *nova*. For the sun is, in effect, a machine, generating its own energy as it radiates. Moreover, we are certain that the store of energy is limited. It is inconceivable that the sun should be a perpetual-motion machine. The universe is running down and our laws of physics indicate that degradation of energy is a one-way process. A hot body spontaneously cools, but a cool body cannot become hot of its own accord amid cooler surroundings.

As will become evident later on, the sun derives its main energy from conversion of hydrogen into helium. When the chief fuel is nearly exhausted, the sun, instead of cooling, may temporarily draw on other very limited energy sources, such as the helium-helium reaction. The temperature of the solar surface would then rise and the earth would become uncomfortably hot, before the inevitable extinction occurred.

Without a more thorough investigation of the solar interior, we may take some reassurance from the sun's past record of steadiness and conclude that any sudden extinction is highly unlikely. In painting the picture of a sunless world, it has not been my intention to frighten. I wished merely to bring out more vividly the importance of the sun and our absolute dependence on it. The sun will probably continue to shine steadily for hundreds of millions of years in the future. Failure of sunlight is a problem not for us but for our remotest posterity.

Elementary calculations enable us to deduce the fact that the sun possesses enough radiation, stored deep in its interior, to provide for 50,000,000 years of output at the present rate. By similar reasoning, we deduce that radiation takes 50,000,000 years to travel from its birthplace out to the edge of the sun, where it can escape into space. This means a veritable snail's pace—only 50 feet per year—for light that will speed at 186,000 miles a second in the vacuum of interplanetary space!

The sun is the direct source of all available energy—with the exception of tides, volcanoes, hot springs, and atomic energy. Water power would fail but for the continued evaporation from the surfaces of lakes and oceans. Winds rely for their maintenance on the inequalities of solar heating as our earth slowly turns on its axis, like a roast on a spit, warming one side while the other cools. The glow from a burning piece of coal is really transformed sunlight. Millions of years ago, the leaves of a primeval forest imprisoned some solar energy. The action of combustion releases this stored radiation.

Even the exceptions listed above may have a solar relation. The sun contributes to the tides directly, and also indirectly to the lunar tides by keeping the oceans warm enough to remain liquid. The internal heat of the earth, which produces volcanism and geysers, although probably caused by various forms of radioactivity, may be in part primordial. The earth and other planets may have sprung from some sort of solar catastrophe. What about radioactive and fissionable atoms, like uranium? Certainly they are not now being produced by natural processes here on the earth. Nor does it appear likely that the sun is making them now. Most probably they represent material left over from the prenatal state of the solar system.

I have already mentioned the stars that suddenly fade or grow more brilliant. [For a full discussion, see Leon Campbell and Luigi Jacchia, *The Story of Variable Stars* (Harvard University Press, Cambridge, 1941).] These are the so-called cataclysmic variables. The brightness of such a star may change 100,000-fold within a few days. As we extend our investigations to large numbers of stars, with refined measures of their light, we discover that many if not the majority of stars are variable. Charles Kingsley's lines,

> Changeless march the stars above;
> Changeless morn succeeds to even;
> And the everlasting hills,
> Changeless, watch the changeless heaven,

may be lovely poetry but they are scarcely good science. Changes of 50 percent or more in the brilliance of a star are not unusual, and variations in excess of 1 percent are common indeed.

The sun itself may also be a variable star, with a range, however, of only 1 or 2 percent. The sun is remarkably steady in its output of light and heat, as well as in its apparent motion around the heavens.

⊙

The ancients early became aware of the sun's constancy, which *Solar* forms the theme of many solar myths. Nearly all primitive people *Mythology* regarded the sun as a person or, in some cases, as a ball of fire carried by someone. The regularity of the sun's behavior, under these circumstances, was somewhat puzzling. Only a slave would perform with such complete dependability. How, then, was the sun or his bearer reduced to servitude? Numerous legends, told by almost every primitive race, were devised to explain the event. The sun, in a large number of these stories, was supposed originally to have been very erratic. Sometimes he hurried too fast on his journey; at other times he dawdled. On occasion he came too close to the earth; often he was too far away. Sometimes he failed to appear at all. Finally, with great difficulties, the sun was caught in a trap or a net, beaten into submission, and, thereafter, performed his duties with absolute regularity.

There are also numerous myths that deal with the sun's temporary disappearance from the sky, to the great consternation of the world. Sometimes the vanishing is voluntary, occasioned by the sun's displeasure with terrestrial wickedness. The return comes only after prayer, sacrifice, and repentance of the wicked. In other legends, the sun is forcibly carried off and imprisoned. His release requires Herculean efforts by some human or animal. Solar eclipses may have inspired some of these myths. Perhaps others refer to periods of darkness caused by bad weather, or a cloud of volcanic smoke, or dust, for in many tales the period of darkness is counted by days or weeks. It is remotely possible that some may refer to a distant period of history when the sun exhibited a variability of considerably greater magnitude than in the present. The wide hold of these legends on the minds of primitive peoples probably arose from the fear that the sun might suddenly and disastrously vanish from the heavens. Primitive peoples did not take the sun for granted!

To relate in detail the various myths would here serve no useful purpose. The personification of the celestial bodies and of the forces of nature in the man-centered world are entirely understandable. But primitive concepts contain little of scientific value. In a way, however, the myths and legends of the ancients are an early type of science. The stories do represent attempts of curious and inquiring minds to find an explanation for puzzling natural phenomena.

The observational data were limited but, nonetheless, carefully studied.

The ancients were particularly concerned with accounting for the sun's return to the eastern horizon during the night. Numerous modes of conveyance were suggested and various routes proposed. The Egyptians had a particularly dramatic story. They pictured the sun being rowed nightly through caverns in the bowels of the earth, where every night he had to fight off.wild beasts and demons who sought to delay him and his attendants.

The Eskimos believed that he was rowed around, just beyond the northern horizon; and cited the occurrence of the rays of the aurora borealis as proof that the sun was traveling in that vicinity. This argument affords an excellent example of "scientific" reasoning by the primitive mind.

Over the greater part of human history, back into ages of whose records we have the merest fragments, we find worship and deifica-

Fig. 1.   Stonehenge, England. A portion of a Druid temple, dedicated to solar worship. The picture shows the Hele Stone, centered beyond the arch, over which the sun rises at the beginning of summer. (Photograph by Charles H. Coles, American Museum of Natural History.)

tion of the sun as one of man's major activities. Sun worship was as natural as it was inevitable. The sun's regularity and beneficial effects on all living things were obvious facts of observation. Worship of the sun and of its terrestrial analogue, fire, followed as a simple recognition of the beneficent and essential qualities of light and heat. Elaborate religious systems arose, based on the very human desire to control or, at least, to influence so powerful a force.

The great hold that sun worship had upon people in early biblical times appears in the numerous scriptural references forbidding worship of the sun. Moses commanded the Israelites: "Take ye therefore good heed . . . lest thou lift up thine eyes unto heaven, and when thou seest the sun, and the moon, and the stars, even all the host of heaven, shouldest be driven to worship them and serve them."

The cross, which, as a religious symbol, far antedates the advent of Christianity, is a solar device, with the arms signifying solar rays. The cross appears in many forms. Often the sun itself is indicated by a circle at the intersection of the beams, as in the so-called Celtic cross. The swastika symbol, with the bent crossarms, is a solar wheel, suggestive of the sun rolling in its annual course through the sky.

⊙

Early Solar
Science

It is interesting to note that so few items of lasting scientific value emerge from all the vast quantity of folklore about the sun. The ancient Babylonians and Chaldeans made numerous astronomical observations, consisting chiefly of determinations of positions of stars and of the paths of the sun and planets, but drew no significant or generalized conclusions from them, with one notable exception. As early as 747 B.C., they had begun tabulation of solar and lunar eclipses. They had some concept of the nature of eclipses and may even have been able to predict them. But it remained for the Greeks to introduce the idea of the universal natural laws, basic to modern science.

The personalization of the sun, stars, and planets was by no means figurative, and the myths were not devised merely for amusement or for their dramatic value. [For a more detailed account of solar

mythology, see W. T. Olcott, *Sun Lore of All Ages* (Putnam, New York, 1914).] Even among the Greeks they were so generally believed that few dared question the truth of the narrative. When, about 434 B.C., Anaxagoras suggested that the sun was a mass of fiery stone as large as the Peloponnesus, rather than Phoebus Apollo, he was arrested on the charge of disputing the established religious dogmas. The eloquence of his friend, Pericles, saved his life, but Anaxagoras was banished from Athens.

The reasoning that led Anaxagoras to his false estimate of the size of the sun was probably scientific, though the following argument is largely conjecture. But we may reasonably assume that if Anaxagoras gave a size for the sun, he also knew, or thought he knew, its distance. Anaxagoras, in common with most men of his time, believed that the earth was flat. His solar distance was apparently inferred from the fact, well known even at that ancient time, that the height of the sun above the horizon varies with the place of observation. Today we recognize the effect as arising from the curvature of the earth's surface. A journey of some 69 miles, 1/360 of the earth's circumference, causes the heavenly bodies to be displaced by 1 degree. Anaxagoras, unaware of the correct explanation of this phenomenon, reasoned otherwise quite logically that the apparent displacement arose from the finiteness of the distances of the heavenly bodies and could thus be used to triangulate the universe (Fig. 3). His argument would consequently lead to a solar distance of about 4,000 miles from the surface of the earth and a solar diameter of 35 miles, only slightly less than that of the southern peninsula of Greece. This result, though erroneous, was nonetheless a marked step forward in man's understanding of the sun.

Anaxagoras, however, was not the first person to contribute to the advance of solar science. It is frequently said that Thales of Miletus had predicted a solar eclipse in the sixth century B.C., but

---

Fig. 2.   Egyptian sun lore. The Egyptian religion included sun worship as one of its tenets. The sun, upper right, sends its life-giving rays to earth-dwellers. Note, also, the flowerlike plants. The rays end in humanlike hands, two of which bear the Egyptian symbol of life, a sort of inverted T with an attached oval. Some writers indicate that the Christian cross, also recognized as a solar symbol, may have evolved from this Egyptian character. (Metropolitan Museum.)

this story is now regarded as apocryphal. However, Thales probably recognized the fact that the moon is a sphere that shines by light reflected from the sun. He announced, moreover, that the heavenly bodies moved according to fixed laws. Thales was shortly followed by Pythagoras, famous for his well-known theorem of the square of the hypotenuse. Pythagorean philosophers gave us our first model of the solar system in which the earth appeared as a planet in motion about a central body. This body, however, was not the sun, but an invisible central fire, whose existence they postulated by a reasoning more mystical than scientific. We do not know for certain the exact views of Pythagoras on this question, since history attributes the foregoing picture to Philolaus, a student of the Pythagorean school who came about 75 years after the master.

The concepts of Thales, Pythagoras, and Anaxagoras were not widely acknowledged as correct. Aristotle presented a more idealized picture of the heavenly bodies. The great philosopher, in the fourth

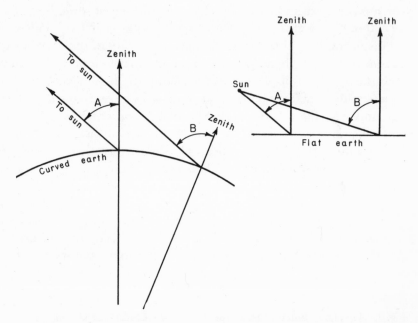

Fig. 3.   The geometry of Anaxagoras. Believing that the earth was flat, Anaxagoras measured angles A and B (*right*) and thought he had triangulated the distance of the sun. Actually, the earth is curved (*left*), so that angles A and B determine the curvature. Because the sun is so far away, the lines pointing to the sun are very nearly parallel.

century B.C., taught that the sun was pure fire and that the stars and planets were also pure and undefiled and quite unearthlike in character. Aristotle's pronouncements, although farther from the truth than those of the earlier natural philosophers, carried greater authority. But with the fall of the Roman empire there was little science of any sort in Europe for nearly a thousand years.

After the revival of learning in Europe, the Aristotelian world picture was generally accepted; in fact, its influence lasted for nearly twenty centuries, until the birth of modern astronomy. Had the views of Aristarchus or of Hipparchus been as widely available and as generally accepted as those of Aristotle, the renaissance in science might well have begun much earlier than it did, for Aristarchus and Hipparchus came much closer to modern concepts of the universe and the workings of the solar system. In the third century B.C., Aristarchus visualized the planets, including the earth, as circling around the sun at the center—essentially the same system that

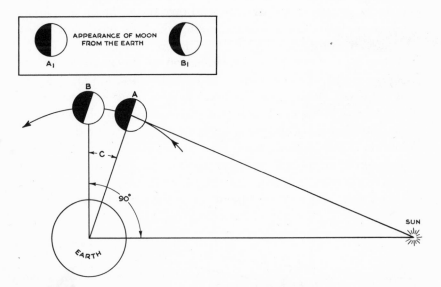

Fig. 4.   Aristarchus' determination of the solar distance. When the moon is at A, we see it exactly at first quarter, as an illuminated semicircle (A₁). When the moon is at B, exactly 90° from the sun, the moon appears slightly gibbous (B₁). The more distant the sun is, the smaller will be the angle C between these two positions. Aristarchus estimated the angle C to be less than 3°; actually it is about 20 times smaller, or about 9'. It takes the moon only 12 minutes to traverse the distance AB, whereas Aristarchus' estimate would have given the time as about 4 hours. (Diagram courtesy of *Popular Science*.)

Copernicus elaborated in A.D. 1543. Aristarchus knew the earth to be a sphere. He further recognized that night and day resulted from the earth's axial rotation. He attempted to measure the true sizes and distances of both the moon and sun. Though he fell far short of deriving their exact values, the methods he used were scientifically sound and failed only because accurate observations were impossible with the crude instruments of his day. Aristarchus judged that the sun was about nineteen times as far away as the moon. He similarly underestimated the distance of the moon, so that his final measure of solar distance came out to be some 720,000 miles—too small by a factor of about 130. Even so, he clearly recognized that the sun was at least as large as the earth. Aristarchus apparently even recognized the stars to be distant suns.

Hipparchus, often called the father of modern astronomy, in the second century B.C. made the best determination of the solar distance of all the ancient astronomers. Although Hipparchus, unfortunately, accepted the geocentric rather than the heliocentric view of the solar system, he measured the size and distance of the moon and knew the approximate diameter of the earth, which he recognized to be a sphere. He also understood the true nature of eclipses.

Hipparchus' method for determination of the lunar and solar distances was completely sound. In the schematic Fig. 5, E represents the earth. At the time of a lunar eclipse, from the curved edge of the shadow of the earth upon the lunar surface, M, he measured accurately the size of the earth's conical shadow at the distance of the moon. Thus, on a scale diagram he could draw circles representing earth and moon and also the line BC, the cross section of the earth's shadow. He then drew the lines BD and CF, just touching the earth, and extended them on to the right of the diagram.

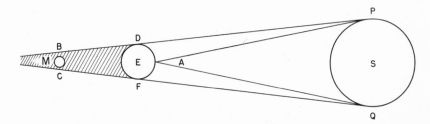

Fig. 5.   Theory of Hipparchus. The diagram (not drawn to scale) shows the circumstances of a total lunar eclipse. The moon, M, lies in the shadow of the earth, E. The circle S represents the sun.

He also knew the angle *A*, subtended by the sun as viewed from the earth. The lines so drawn intersected the previous lines at *P* and *Q*, so that he could measure both the diameter and distance of the sun, *S*.

The weakness of this method—no fault of Hipparchus'—lies in the fact that *ES* is so very much greater than *EM* that the lines *BDP* and *AP* intersect at so small an angle that the point *P* is difficult to determine. Hipparchus apparently realized these limitations and used the study to redetermine the distance of the moon, which he got quite accurately. He then employed Aristarchus' old result that the sun was at least nineteen times farther away than the moon to recalculate a new limit for the size and distance of the sun. Thus Hipparchus succeeded in showing that the sun's diameter was at least seven times greater than that of the earth.

Hipparchus' estimate of the solar size and distance remained substantially unchallenged until about 1620, when Kepler arbitrarily increased them by a factor of 3. The reasons for the change are ambiguous. In part the changes were dictated by observations of his great master, Tycho, but numerology (an unfortunate predilection of Kepler) may have played a part. About 1650 Riccioli put in an additional factor of 2. Although Vendelinus had argued for a factor of 4, his value was not accepted.

In 1672, the well-known astronomer Cassini set a lower limit, which proved to be within 10 percent of the modern figure, a result essentially confirmed by observations of the transits of Venus in 1761 and 1769. After an enormous interval of time, scientists finally surpassed the results of Hipparchus. What other scientist's results, sound in themselves, have stood the test for 1700 years? Meanwhile, of course, had come Galileo, with the invention of the telescope and the discovery of sunspots—but that story must wait, for the moment, while we consider further the important question of the distance of the sun from the earth, upon which figure depends all of our knowledge of dimensions in the solar system and in the universe of stars as well.

⊙

Distance of the Sun

Many methods for finding the solar distance are available, some better than others, so that I shall discuss only a few. Since the earth's orbit is slightly elliptical, the actual distance of the sun varies from day to day. Hence we choose for the value half of the major axis of the ellipse (Fig. 6), and call it the sun's mean distance. In theory the determination is quite simple, based on elementary geometry and trigonometry of the sort surveyors use in measuring the distance to some inaccessible station, let us say across a lake. For example, suppose that a surveyor at *A* wishes to find the dimension *AB* (Fig. 7). He chooses some convenient third station *C*, chains the distance *AC*, and measures the angles *A* and *C* with a transit instrument. These figures, incidentally, enable him to evaluate the angle *B*, since the three angles of a triangle must sum to 180°. Then, by trigonometry, or with a geometrical diagram and simple ratios, he can compute the desired distance *AB*.

We carry this method over directly to astronomical problems. Let *A* and *C* represent two stations on the surface of the earth and *B* the center of the celestial object whose distance is desired. Assume that surveyors have already measured the distance *AC*. Observations give the angles *A*, *C*, and *B*, and calculation ultimately leads to the required distance. The longer the baseline *AC*, the more accurate

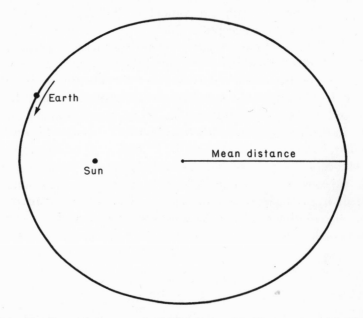

Fig. 6.   Earth's orbit seen as an ellipse (schematic). The sun is at one focus of the ellipse.

the determination of distance, as explained in the legend of Fig. 8.

This method works very well for the nearer bodies. Let us suppose, for example, that an observer at A (Fig. 9) notes the apparent position of the moon against the background of distant stars, and that an observer at C, on the opposite side of the earth, does the same. The moon, as observed from the two sites, will occupy positions differing by almost 2°, or about four times its own diameter. Half the angle ABC, indicated by the arrow, is called the parallax, by definition. The parallax of the moon is 57′ of arc, corresponding to a distance slightly less than a quarter of a million miles. The angle is measurable by even relatively crude instruments and, in consequence, the ancients had a fairly good idea of the lunar distance.

This method, fundamental as it is, fails for two reasons when we apply it to the sun. One is that the angle at B is 400 times smaller than it is for the moon; we thus require much greater accuracy of measurement. But the great obstacle is that the stellar background is invisible during the day, so that our reference points disappear. We must, therefore, have recourse to indirect methods.

Without measuring a single celestial distance, the astronomer finds it a relatively simple task to draw a map of the solar system to scale. As one of the simpler examples, he may note that Venus never gets farther from the sun than 45°. The conditions are about as shown in Fig. 10, and one easily calculates, from the properties of a 45° triangle, that the distance B is about 0.7 of distance A. Hence, if we draw a map where 1 inch represents the radius of the earth's orbit, 0.7 inch will represent the distance of Venus from the

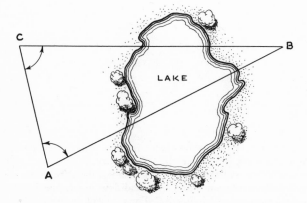

Fig. 7.   The surveyor's triangulation. (*Popular Science.*)

sun. The orbits of all the other planets may be similarly drawn, without much greater complication. The problem, now, is to determine the scale of our map. We can fix this scale from measurements of the length of any line upon the chart, say the distance of Venus from the earth, quite as well as from direct measurements of the solar distance.

This indirect procedure greatly simplifies the determination. First of all, Venus, when nearest the earth, is far closer than the sun. On occasions, Venus comes directly between the earth and sun, producing what we call a "transit of Venus." At those times the astronomer uses the apparent position of the planet, viewed against the solar disk, to determine the distance of the sun. The face of the sun serves in place of the reference background of stars, although we must make due allowance for the fact that the distance of the sun

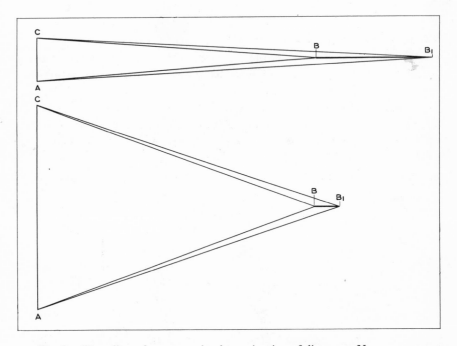

Fig. 8.   The effect of errors on the determination of distances. No measurement of an angle is absolutely precise. In the lower figure, the two lines radiating from A and C represent the limits of error. Consequently, the uncertainty of determination places the object between B and B_1. Conditions are similar for the upper diagram, except for the smaller baseline AC. The range of uncertainty in the position, BB_1, is correspondingly greater. (*Popular Science*.)

is finite, where that of the stars is practically infinite. Observations of Venus in solar transit, taken from two separated stations on the surface of the earth, constitute one important method for fixing the solar parallax. For many years, results thus derived were the standard. Unfortunately, however, transits of Venus are rare phenomena indeed. A pair occurred in 1874 and 1882, but we shall have no others until June 8, 2004 and June 6, 2012.

Mars has also been used for determinations, and likewise the asteroid, or tiny planet, Eros, which comes, at the most favorable times, within 14,000,000 miles of the earth, more than two and a half times closer than Mars ever comes. This circumstance, coupled with the fact that Eros looks like a star, enables us to measure its position accurately against the reference background of stars. Observations of Eros provide one of the best methods devised to date

Fig. 9.   Triangulation of the moon.

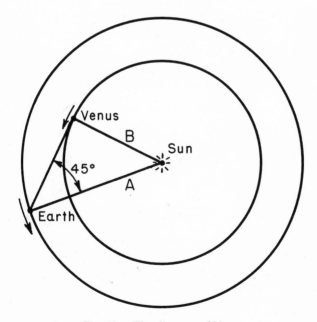

Fig. 10.   The distance of Venus.

for determining the distance of the sun. Observations of Mars were the basis of the previously mentioned attempt by Cassini in 1672 to measure the solar parallax. In Fig. 11, the radius of the earth's orbit can be calculated, once the distance *ab* or *cd* is known.

Still other methods exist, one of them simplicity itself. As I shall explain later on, the spectroscope affords a ready means of measuring the velocity of motion of any object toward or away from the observer. In Fig. 12, let the circle at the right represent the earth's orbit, with *A, B,* and *C* three possible positions of the earth. Now suppose that we measure spectroscopically the radial velocities of a number of stars, *D,* that lie opposite to the sun when the earth is at *B*. Their speeds will average very nearly to zero, as one would expect, for their own motions are generally at random. Some are plus and some minus, indicating velocities away from and toward the observer, who is at the moment moving crosswise to the line of sight. The spectroscope records the velocities only *in* the line of sight. If, now, we compare observations taken three months earlier, when the earth was at *A*, we shall find these stars apparently rushing toward the earth with a speed of about 18.5 miles per second. Sim-

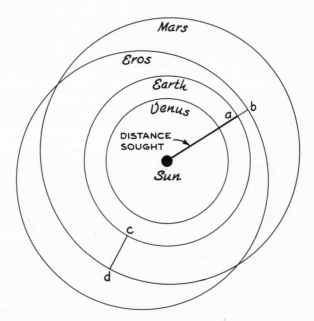

Fig. 11.   Orbits of the inner planets.

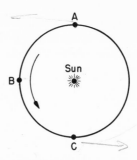

Fig. 12.   Spectroscopic determination of the solar distance.

ilarly, when the earth is at *C,* the average velocity of recession of these stars is again 18.5 miles per second. (See also Fig. 33.)

This effect, certainly, must be ascribed not to the stars but to the earth itself, moving in its orbit with a speed of 18.5 miles per second. If, then, we multiply 18.5 by the number of seconds in a year, we obtain the distance traveled by the earth in the interval, or the circumference of the terrestrial orbit. Finally, dividing by 2π ($\pi =$ 3.1416), we arrive at the radius of the earth's orbit.

Dr. Gerald Clemence, Director of the Nautical Almanac of the U.S. Naval Observatory, advises me that results obtained by the so-called "dynamical method" are preferable to those from either trigonometric or spectroscopic procedures. The dynamical parallax depends on a thorough study of the motions of the planets and of our moon in particular. The continual gravitational tug of war between the sun and the earth for the moon gives an indirect measure of the solar distance that is considered to be more precise than any given by the direct methods. According to Rabe, the solar parallax is 8″.7984, corresponding to a distance of 92,914,000 miles or 149,530,000 kilometers, with a probable error of about one part in 20,000, or roughly 40,000 miles. The magnitude of this figure is indeed difficult to grasp. An aviator flying continuously at a speed of 500 miles an hour would require 21 years to cover an equal distance. Even at the speed of an orbiting earth satellite, 4½ miles per second, he would need more than 8 months to reach the sun.

We can readily calculate the size of the sun, once we know the distance. The angle in the heavens from one edge of the sun's disk to the other is slightly over half a degree. A cigarette, ¼ inch in diameter, pointed toward the sun and held 26¾ inches from the eye, will just eclipse the disk. The diameter of the cigarette is 1/107 of its distance from the eye. Similarly, the diameter of the sun will be 1/107 of its distance, or 865,000 miles.

⊙

Mass of the Sun The solar mass, that is, the quantity of matter that the sun contains, is a little more difficult to evaluate. As the first step in the process, we must determine how strong the force of gravitation is. Figure 13 illustrates the principle. Two equal and known masses A and B hang from the opposite ends of the beam of a very sensitive balance. A third body, C, also of known mass, is placed beneath A. The mutual attraction of A and C causes A to swing downward until a fourth, very small but measurable, mass D, added to B, restores the equilibrium. Since the attraction of the entire earth on D equals that of C upon A, we can deduce the mass of the earth as 6.59×10^{21} tons. Numerous alternative methods exist, some more accurate than the one I have just described, for ascertaining the mass of the earth, but all go back to this same fundamental principle.

The earth swings in its annual orbit almost as if an invisible string connected it to the sun. Indeed, the pull of gravity simulates the tension of a string, so that the earth, instead of flying off on a straight line, as it would if the string broke, moves toward the sun. In fact, one may say that the earth "falls around" the sun, the departure of its orbit from a straight line being less than ⅛ inch in the course of a second (Fig. 14). Now at the surface of the earth a body falls 16.1 feet in the first second, as has been known since the days of

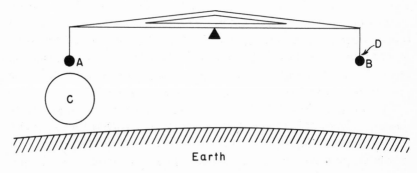

Fig. 13.   Determination of the mass of the earth.

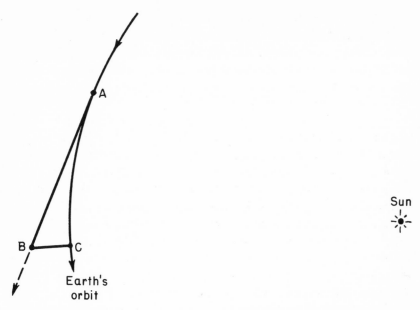

Fig. 14.   The effect of gravity upon the earth's motion. If AC represents the distance the earth travels in 1 second (18.5 miles), the departure BC from a straight line is only ⅛ inch.

Galileo. These distances, ⅛ inch and 16.1 feet, are directly proportional to the respective gravitational accelerations, forces acting on a unit mass, caused by the sun at the distance of the earth and by the earth at its surface. Hence, knowing that the gravitational acceleration is directly proportional to the mass and inversely proportional to the square of the distance from the center of a sphere, we readily calculate that the sun has a mass 329,390 times that of the earth. And, introducing the mass of the earth from the previously described experiment, we find that the sun has a mass of 2.24×10^{27} tons. (Written in full, this number is 2,240,000,000,000,000,000,000,-000,000 tons. The abbreviated notation will be used henceforth. The exponent 27 determines the number—1 followed by 27 ciphers—by which the first figure must be multiplied to give the final result. Thus 10^2 is a hundred; 10^6, a million. Very small numbers are indicated by negative exponents. Thus $6.32 \times 10^{-6} = 6.32/10^6 = 6.32/1,000,-000 = 0.00000632$.)

We are now in position to calculate the average density of the sun, that is, its mass divided by that of an equal volume of water. Since one cubic centimeter of water weighs one gram, we have merely to divide the solar mass (in grams) by its volume (in cubic centimeters). The result proves to be 1.42. An *average* sample of the sun would weigh about as much as a lump of soft coal of equal size. The inner solar layers are more highly compressed and may attain densities as great as 77 at the center, about 10 times greater than that of steel. The outer portion is correspondingly less dense than the average. The structural details of these surface layers will be discussed in the next chapter.

2

A Panoramic Sunscape

Have you ever seen the sun? Really seen it, in its intricate detail? We are all aware, of course, that somewhere in the sky is a brilliant source of light, which hurts our eyes if we accidentally look at it directly. But probably you have never really seen it. When the heavens are partially overcast or when the sun is close to the horizon, we see the sun's circular disk, but little else of detail. Occasionally, a sunspot group, when extra large, may be visible to the naked eye. The ancient Chinese annals contain many references to spotted areas of the solar disk. Groups of this magnitude are fairly uncommon, however, and most spots can be detected only with telescopic aid.

Never look at the sun without some protection for the eye! Very dark glasses, a piece of heavily smoked glass, and fogged photographic films are most commonly used. Two Polaroid filters, superimposed, are ideal, because rotation of one member with respect to the other allows the observer to select exactly the desired intensity of transmitted light.

Any filter tends to change the color of solar radiation. For that reason the student may wish to adopt a simple and effective filterless method. Get two or three small pieces of plate glass or unsilvered mirror. Set one of these pieces on a chair, facing the sun. The reflected image will still be uncomfortably bright. But the reflection of this image in a second or even a third piece of glass will generally be faint enough for clear viewing. An unsilvered *pentaprism* (Fig. 15), which provides for loss of sunlight at two of its faces, is even more convenient and may be adapted to telescopic use.

Such an experiment is well worth performing, for it gives a very clear idea of the sun's apparent diameter, which is much smaller than the average person would estimate from casual observation. In addition, it emphasizes the silvery color of sunlight. Most persons have seen the solar disk when the sun is near the horizon, reddened by absorption in the long path of intervening air. They therefore think that the sun is orange, or, perhaps, golden. Actually, sunlight is white, or very nearly so.

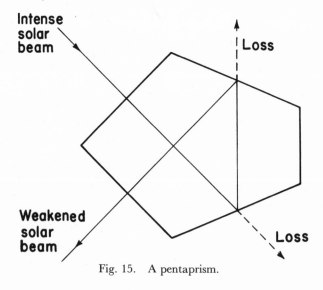

Fig. 15.   A pentaprism.

A pinhole will serve as a projector, especially when used to form an image of the sun within a darkened room. A 1-inch image will appear on a piece of white paper held 10 feet away from the pinhole. Large spot groups should be clearly visible. But, if any simple optical aid, like a telescope, field glasses, or even modest opera glasses, is available, do not bother with the relatively ineffective pinhole. With dark glass placed before the lens, one may easily view the sun directly and see its surface characteristics. Take great care with the dark glass, however, for if you look at the sun directly, without such protection, serious damage to your eye can result.

Astronomers frequently employ still another method of observation. Focus one of the instruments described above upon a white screen held a foot or two behind the eyepiece. The enlarged solar image, thus projected, can be examined directly. A second screen, with a hole to admit the objective, should be placed at the front end of the telescope, to cast a shadow upon the screen.

The amateur may readily photograph the sun if he possesses a small telescope and a camera that permits ground-glass focusing, either direct or reflex. Attach the camera directly to the eyepiece of the telescope. Or set the camera on a tripod behind the telescope eyepiece, directly in line with the inclined tube. If possible, remove the camera lens. Move the eyepiece back and forth to focus the image on the ground glass. To reduce the brightness of the sun's image, insert a glass filter (dark yellow or orange). Then take the picture in the ordinary fashion. A steel-bladed shutter is better than one made of hard rubber, because the latter may melt or burn in the concentrated solar rays. As a precautionary measure, when the shutter is closed, keep the objective covered except at the moment of taking the picture.

If your camera does not have a device for visual focusing, you may still take pictures, though you will then have to determine the precise focus by trial and error. Attach the camera, with lens removed, to the telescope eyepiece, which must have a scale so that you may record and repeat the correct focus setting.

Photographs of the outfit of the late Rev. William M. Kearons, of Fall River, Massachusetts, appear in Fig. 16. Several pictures taken with this solar camera are reproduced elsewhere in this book. Many amateurs regularly send their daily photographs to Zurich, Switzerland, the center for the preservation of such records. A sec-

Fig. 16. (*a*) Rev. William M. Kearons, with small telescope fitted for solar photography. (*b*) Alternative mounting with camera on tripod.

tion of the American Association of Variable Star Observers also devotes attention to solar phenomena.

Some sort of telescope is absolutely necessary for the student who wishes to observe at first hand even the more obvious features of the solar surface. He may construct a simple but nevertheless effective instrument easily and cheaply from a couple of lenses and a cardboard mailing tube. For the primary image-forming lens, known as the objective, use a simple convex spectacle lens, which costs about one dollar. The power of 1.5 diopters, which signifies that the lens will bring the sun to focus at a distance of 1/1.5 meters, or about 26 inches, is fairly convenient. Mount the objective permanently in one end of the tube. For an eyepiece, a two-lens hand magnifier having a focal length of about 1 inch is ideal. Set these lenses into one end of a second and smaller tube that slides within the first. Figure 17 suggests one form of construction.

The principle of the telescope is clearly shown by this figure. The objective forms an inverted image of the sun, which one views by means of the magnifying eyepiece. In the strict sense, one does not look through a telescope, but only at the image formed by the primary lens. This type of instrument, which employs a lens objective to focus the rays, is known technically as a *refractor*. In the simple form described it is none too satisfactory, although it will show sunspots and the craters of the moon quite clearly. Difficulty arises since prismatic effects of the lens cause light rays of different colors to bend at different angles, the blue more than the red. Hence, as we focus our eyepiece upon any one shade, the other colors will be slightly out of focus.

One can minimize this difficulty by using an objective that consists of two or more components of different kinds of glass. Or one may employ the reflecting telescope, with its concave mirror, instead of a convex lens, to focus the rays. The mirror is ordinarily silvered or coated with evaporated aluminum on the front face. But for solar work, the mirror is often left uncoated, to reduce the amount of light in the image of the sun. [For a general discussion of optical instruments, see G. R. P. Miczaika and W. M. Sinton, *Tools of the Astronomer* (Harvard University Press, Cambridge, in press).] Remember, whatever method you use, always protect your eye when looking at the sun directly, with or without a telescope. One may use dark or smoked glass, or alternatively a *solar eyepiece,* which wastes a large part of the unwanted light at reflection from non-silvered surfaces, as with the pentaprism previously described.

However the sun is observed, one feature should be entirely obvious: the solar disk is by no means uniformly bright. The intensity is greatest at the center and diminishes toward the limb, as astronomers term the edge of the sun. This phenomenon, known technically as limb darkening, arises because the solar surface, instead of being solid and sharply bounded, consists of a layer of gas. Near the edge of the sun, where we look tangentially into this solar atmosphere, the layers we see are cooler and hence appear less bright than the deeper levels we view near the center of the disk.

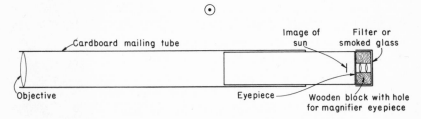

Fig. 17.   Construction of a simple refracting telescope.

Sunspots   The widely publicized phenomenon of sunspots is the most obvious variable feature of the sun. When Galileo turned his telescope sunward he was undoubtedly surprised to find dark areas here and there against the shining solar surface. A few weeks of observation sufficed to convince him that these spots were truly part of the sun and not planets or other solid bodies seen in projection. Galileo found the sun to be rotating, and that a few days less than a month were required for a complete turn on its axis. Recognizing the far-reaching significance of his observation, which was so at variance with the philosophy of his time, Galileo hesitated to publish his results. Meanwhile others, among them Fabricius and Scheiner, discovered the phenomenon independently, although the latter argued that the spots could not actually be on the sun.

At last Galileo spoke, simply and effectively. "Repeated observations have finally convinced me that these spots are substances on the surface of the solar body where they are continuously produced and where they are also dissolved, some in shorter and others in longer periods. And by the rotation of the sun, which completes its period in about a lunar month, they are carried around the sun, an important occurrence in itself and still more so for its significance."

The individual spots proved to be temporary phenomena. The smaller ones are very short-lived, lasting but a day or so. The larger ones usually persist for some days. Relatively few remain visible

Fig. 18.   Giant sunspot group, 7 April 1947. (Mount Wilson Observatory.)

Fig. 19.   Drawing of a sunspot. (Langley.)

long enough for solar rotation to bring them into view a second time.

High telescopic magnification reveals the detailed structure of a sunspot: the dark central core, known as the *umbra*, rimmed by the delicate filaments of a less dark region, the *penumbra*. Sunspots are never exactly round. The edge of the umbra is usually jagged, with the outline of the penumbra roughly parallel to it. In places, the filaments of the penumbra project well beyond the general outline of the spot, like little patches of hay spilled around the edges of a haystack.

Occasionally, spots occur singly. More often they appear in pairs, or in complex patches. A representative group ordinarily contains two spots of major dimensions, with a number of smaller spots irregularly distributed in the neighborhood. A large sunspot is often the scene of great activity. At times, a brilliant bridge may form and cut the spot in two. The entire character of the phenomenon may change markedly in the course of a few minutes. Then, after days or weeks, the spot may disappear, leaving bright veins as the last trace of the disturbance. This pattern of bright patches associated with spots we call *faculae*.

The original discovery of sunspots presented a difficult philosophical problem as to their nature. I have already pointed out that most primitive races had regarded the sun as a god. Aristotle, whose views on most subjects were unquestioningly accepted in Galileo's time, had stated that the sun was a ball of *pure fire*. I italicize both words because the emphasis was fully as much on the first word as on the second. Consequently, a large number of people regarded the alleged existence of spots as impeaching the pure character of the sun, because spots and purity could not be reconciled. Many persons refused to look through the telescope, lest they, too, become bewitched and see this defilement that did not accord with the teachings of Aristotle, prince of philosophers. The invention of the telescope and the discoveries made with its aid contributed importantly to weaken the medieval devotion to authority, and to promoting the experimental and observational approach to reality which made possible the rise of modern Western civilization.

For sunspots proved to be real phenomena. Galileo regarded them as clouds floating in the solar atmosphere, and, in this view, he was less far from the truth than many who followed him. Others argued that the spots were mountains projecting above the luminous clouds. Many thought of them as the product of some sort of volcanism.

For many years, astronomers believed that spots were storm areas —a conclusion drawn from analogy with the earth's atmosphere. Gases in high-pressure areas tend to flow toward regions of low pressure. The resulting expansion of the gases cools them. On earth the moisture-laden air, thus cooled, can hold less of its water vapor. Condensation and precipitation result. On the sun the gases are everywhere far too hot for condensation. Even the heavy metals occur in vaporous form. But the cooler areas manifest themselves visibly by radiating less energy than the surrounding regions. Thus astronomers concluded that the dark, cooler areas were probably solar storms.

However, recent evidence, largely of a theoretical nature, indicates that the spots are not storm areas after all. We had failed to realize that the intense magnetic fields known to exist in the spots impart some degree of rigidity to the hot gases, so that the violent winds associated with intense storminess simply cannot occur. The spots, on the contrary, are islands of intense calm in the vast stormy solar atmosphere. Indeed, their coolness itself derives from the fact that no heat transport can take place in them by convection. To compensate for the lower convection in spots, the areas immediately outside of the spot have increased convection, with turbulence that comes much closer to the surface than usual.

We find spots in all varieties of sizes. The smallest ones, which are usually called "pores," a few hundred miles or so across, are the most numerous. Very likely even smaller ones occur, but they are too tiny to be seen through a telescope. At the other end of the scale, we often find spots whose diameters, including the penumbra, may measure 20,000 miles or more. A large double group may extend 100,000 miles or more, with an area of several thousand million square miles. The largest spot group on record, which appeared in April 1947, covered more than 1 percent of the area of the apparent solar disk. Its total area was about 6000 million square miles. This spot group was large enough to contain about 100 earths. Any spot larger than 25,000 miles in diameter is readily discernible to the naked eye.

⊙

If one follows the appearance of the sun from day to day and year to year, keeping records of the numbers and areas of spot groups, he will find a gradual and fairly regular rise and fall of solar spottedness. Schwabe, of Dessau, in 1843, first suggested the existence of periodicity in sunspots from 17 years of his own careful observations. This suggestion received little notice until 1851, when Humboldt included 25 years of Schwabe's data in his *Kosmos*. Later, R. Wolf, of Zurich, found an unpublished manuscript dated 1776, written by the Danish scientist Horrebow, which noted the possibility of a periodic variation in the sunspots. Wolf succeeded in uncovering many observations by the earlier astronomers and was thus able to extend our knowledge of sunspot numbers back to the invention of the telescope in 1610, with a practically complete record from 1745.

The average spacing between successive maxima is 11.2 years, but the relation is by no means exact. A maximum occurred in 1788 and the next was not reached until 16 years later, in 1804. On the other hand, the maximum in late 1829 was quickly followed

Solar Variability

by one in early 1837, an interval of only about 7.5 years. The problem of sunspot variation is far too complex for full discussion in this introductory chapter. We shall return to these interesting questions later on.

⊙

Granulation and

Faculae

Sunspots were first discovered because their sharp contrast makes them conspicuous, even in a small imperfect telescope. Much higher power and better instruments are required to disclose the structure of the shining surface between the spots. This area presents a granular appearance often referred to as "rice grains." This beautiful simile must be attributed to the astronomer who asserted that the granulations look like "rice grains floating in a bowl of soup." The description is suggestive, even though it displays the scientist's ignorance of the indisputable culinary fact that rice grains sink to the bottom of the bowl.

The solar rice grains, tiny brilliant patches several hundred miles across, standing out against the darker background, appear mostly near the center of the disk. Photographs show that the individual grains are short-lived. Within the course of several minutes the entire character of the pattern may change completely. The rapidity of the variations indicates the turbulent state of the sun's atmosphere. On the earth we refer to a 90-mile-an-hour wind as a hurricane. On the sun, atmospheric motion of that speed would be counted as mere stagnation, for velocities upward of 90 miles a second are not

Fig. 20.   Faculae: bright streaks around a sunspot. (Meudon Observatory.)

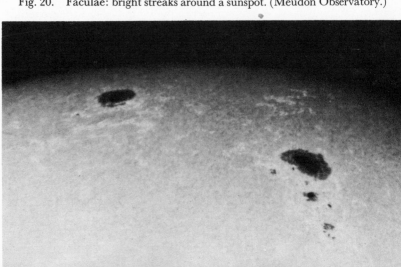

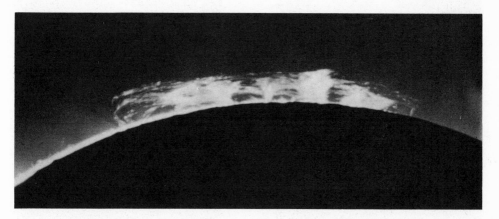

Fig. 21.   Prominences at the solar eclipse of 1905. (Lick Observatory.)

uncommon, especially in the upper atmospheric levels. The rice grains appear to be waves on the stormy ocean of the sun's surface. These waves arise from convection, the well-known tendency of hot gases to rise and cold gases to sink.

As previously noted, magnetism inhibits convection within the spots. Hence greater than normal convection tends to occur in the regions just outside the dark areas, with the result that around sunspots, we encounter great patches of bright material called *faculae* ("little torches"). These faculae form a veined network much larger than the dark spots. Also they seem to be more permanent than the spots, probably because magnetic fields can cause them even when the fields lie concealed below the surface. Presumably they are elevations of a sort, temporary mountains produced by gas flowing through a certain region of the solar atmosphere. Faculae are most conspicuous near the limb.

The turbulence of the sun's atmosphere is shown most strikingly, perhaps, by the great clouds of gas that occasionally extend above the solar limb. These clouds, called *prominences* or protuberances, have the appearance of enormous flames. Although the clouds are luminous, they are not flames in the ordinary sense, because they do not owe their luminosity to combustion. The sun is not "on fire."

Prominences

211600 U.T.

211845

 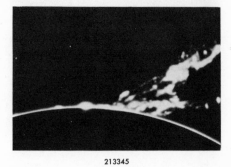

212115   213345

Fig. 22.   Four stages in the development of a solar limb event, 10 February 1956. The notation 211600 U. T. means that the time was 21ʰ16ᵐ00ˢ Universal or Greenwich time. (Sacramento Peak Observatory.)

Prominences are visible only with very special equipment, to be described later on. But at the time of total solar eclipse, when the moon blots out the brilliant disk of the sun, prominences appear as elevated rosy patches of cloud, sometimes floating high above the solar surface.

The motions and forms of prominences are subjects for special investigation. Some of the clouds are of truly eruptive character, shot upward with explosive violence. Others remain suspended like fleecy summer clouds, showing evidence of internal motion, and all but detached from the solar surface. Some form swiftly at high levels, with no obvious source, and rain luminous streams sunward in graceful curves. Still others, usually associated with sunspots, shoot out like a ribbon of flame and then withdraw as if the sun had momentarily extended a snakelike tongue. Prominences and sunspot activity are clearly related. The boiling, bubbling convection of the atmosphere assisted by the forces of electromagnetic currents tends to expel matter. Local magnetic fields determine the paths taken by the flowing gas. Shock waves, resulting from disturbances in the outer borders of sunspots, may be responsible for various types of prominence formations.

⊙

Prominences, particularly the high-ascending ones, move through *The Corona*
a region of space that is still far from being a perfect vacuum. The
sun is completely enveloped in an aura of tenuous atmosphere. The
atoms of this solar fringe, which comprises the *corona,* emit a feeble
radiation that the bright glare of the sun and surrounding sky
light ordinarily conceal entirely. Only when the sun is completely
eclipsed by the moon does the corona come into view. We then
perceive it as a delicate system of rays or "petals," with a distinct
structural pattern of filamentary streamers. I shall later describe
the special instrumental techniques devised for detection and study
of the corona outside of eclipse.

The coronal streamers extend millions of miles out into space.
On occasion, they may even brush the earth, producing magnetic
storms and brilliant auroral displays in transit. Indeed, recent evidence points toward the possibility that the corona may normally
extend out beyond the earth, so that our planet is actually immersed
in the tenuous outer layers of the sun.

We shall eventually discuss these interesting problems in detail.
For the present, however, we shall be content with noting that the
corona, like sunspots and prominences, also depends for its form
and structure upon the condition of solar activity. At sunspot maximum, the corona presents a wind-blown aspect, with streamers
stretching irregularly in every direction. At sunspot minimum, the
corona appears carefully combed, with a neat part at the poles and
long hairy streamers stretching out from the equatorial regions.

⊙

The rotation of the sun, already mentioned, presents an interesting *Solar Rotation*
puzzle. Its rotation is not like that of a solid body. The sun's equatorial regions make the circuit in less time than do the intermediate
latitudes, a conclusion drawn from the fact that spots near the
equator move ahead of those nearer the poles. The reason for this
equatorial acceleration is unknown. The planets Jupiter and Saturn
show a somewhat similar behavior.

Determination of the direction of the axis of rotation is a simple
matter. If the axis were perpendicular to the plane of the earth's
orbit, spots would always seem to move in straight-line paths across
the disk. Actually, one observes this sort of motion only about June

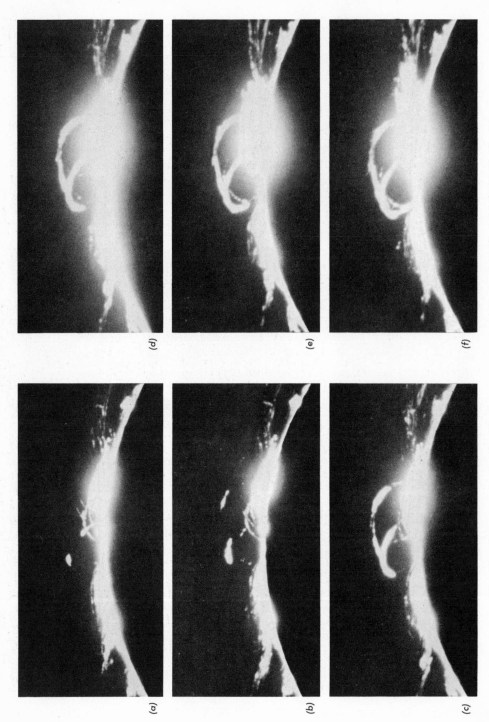

Fig. 23.   Downward growth of loops near a sunspot, 14 April 1947: (a) 16ʰ16ᵐ U. T.; (b) 16ʰ26ᵐ; (c) 16ʰ46ᵐ; (d) 16ʰ51ᵐ; (e) 16ʰ56ᵐ; (f) 17ʰ01ᵐ. (High Altitude Observatory, Climax.)

Fig. 24.   Solar corona, 1936. (Photograph by Irvine C. Gardner of the National Bureau of Standards, copyright National Geographic Society.)

6 or December 6. At other times spots travel in curved paths, the maximum of curvature occurring at the intermediate dates of March 7 and September 8. The appearance is shown in Fig. 25. The curvature is not great but, when carefully measured, it fixes the inclination of the sun's axis at about 7° to the perpendicular of the earth's orbit plane. From June to December, we can see the sun's north pole, and during the other half of the year, the south pole.

A point on the sun's equator completes a rotation in 25 days. Its speed is some 4500 miles an hour, about four times the velocity of a similar point on the earth's equator. The earth departs measurably from a spherical form. Its polar diameter is 27 miles less than the equatorial, because of the centrifugal force developed by its rotation. The question is: can we observe a similar polar flattening for the sun?

The question is not an easy one to answer, because the sun's departure from sphericity, if any, is very small. True, observers have found differences of the order of 0.01 percent, about 100 miles. But the measures are as likely as not to indicate a polar diameter greater than the equatorial. The difficulty lies in the observations themselves. Light from the lower edge of the sun's disk has to traverse a slightly greater thickness of the earth's atmosphere than does light from the upper edge. This difference of path, minute though it is, nevertheless introduces errors, with uncertain corrections. We feel sure that some flattening must exist, but we shall have to devise new methods of observation before we can determine its magnitude.

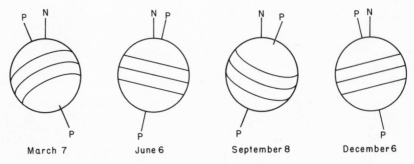

March 7   June 6   September 8   December 6

Fig. 25. Inclination of the sun's axis of rotation. The line PP represents the axis; N is the north point on the solar disk.

The equatorial diameter is much simpler to measure than the polar. At noon, the east and west edges of the sun have the same altitude above the horizon. Consequently, the light paths through the earth's atmosphere are essentially identical, and the corrections are automatically eliminated. We do not have to point our telescope first to one edge and then to the opposite edge of the sun and measure the angle we have turned the instrument through. Instead, we direct the telescope toward the meridian and wait for the sun to cross its field of view. At the moment the preceding limb touches the crosswire we punch a telegraph key that actuates a recording clock, and punch the key again, some 2 minutes later, as the following limb drifts past the index mark. The telescope, meanwhile, is stationary; hence we easily calculate the diameter from the recorded time elapsed between the two contacts.

It is usually much easier to measure time than angles. To achieve an accuracy of 0.01 percent, we have only to record the duration of the transit to about 0.01 second. Since short time intervals may be fixed with still greater precision, if necessary, this part of the study presents no serious difficulty. The earth's atmosphere makes all of the trouble. For, unless the air is exceptionally clear and steady, the edge of the sun is wavy and indefinite, not sharp and well marked as the study requires. Even when the atmosphere is not moving, any trace of haze serves to make the sun appear larger than it really is. This effect, known as *irradiation,* may amount to as much as 0.1 percent.

Long series of observations seem to indicate, in spite of the numerous difficulties, actual variation in the equatorial solar diameter. But the direction of the change is controversial. Secchi and Rosa concluded that the sun is larger during sunspot *minimum.* Meyermann, on the other hand, from a reduction of older heliometer data, recently obtained just the contrary result—the sun is larger around sunspot maximum. In view of the enormous forces of solar activity, a small change in the surface layers should not be the least disquieting. However, the problem obviously needs further study by the best of modern methods. In passing, we may note that many of the stars which vary in brightness also show pulsations in diameter, of 10 percent or more.

I have tried in this chapter to sketch a general picture of the solar panorama: the activity of spots, the mad hurly-burly of the atmosphere, and the superexplosive activity of prominences. I have painted rapidly, with a broad brush, leaving the details to later chapters. Here and there I have hinted at difficulties to come, at questions yet unsolved.

The astronomer's interest in solar problems is enhanced by the fact that the panorama is ever changing. Sunspots come and go. Prominences form, become active, and disappear. The regular march of solar rotation continually brings new features into view. And if the laws governing the motions of the solar atmosphere are still a puzzle to our minds, we ever hope to discover some new principles that will reveal order in the apparent chaos reigning in the sun's outer layers.

The few regularities we have already found, such as the periodic rise and fall of sunspot numbers, encourage us to continue our efforts. We believe in the universal applicability of physical laws. We further expect that everything in nature has an explanation in terms of these laws. If we find points of disagreement, we revise our concept of the law. In many cases the phenomena are so complex that we have not yet understood the ways in which the laws are operating. But, in general, the scientist denies all miracles—except, perhaps, the one that started the universe on its way.

3

Light, Atoms, and Test Tubes

Light is fundamental to the existence of astronomy. One recalls that, in the story by H. G. Wells, "The Country of the Blind," the sightless inhabitants were entirely unaware of the existence of the heavenly bodies. They could, of course, distinguish between night and day, because of the change in temperature. But the blind folk had no concept of the nature of the universe and only theorized that the vault of heaven was smooth to the touch.

Mere blindness, however, does not necessarily eliminate all possibility of finding out something about the universe. There are numerous other ways for exploring the sky. The skin is an insensitive receiver compared with the human eye. But it also can detect solar radiation. And other more efficient physical devices exist: the thermocouple, the bolometer, the photocell, all of which may be readily adapted to tactual or aural recording. With the aid of electrotyping, one could even subject photographs to examination with the finger tips.

The modern astronomer is as much concerned with light he cannot see as with visible radiation. Beyond either end of the rainbow of colors that the eye can perceive lies an even greater variety of shades to which our eyes are insensitive: radiation in the infrared and ultraviolet, below the red and beyond the violet. In recent years we have become aware of even longer waves, radio energy of cosmic origin. We call this entire array of radiation, visible and invisible, the *spectrum*.

The great astronomer, Sir William Herschel, in 1800, performed the experiment of placing a thermometer just outside the visible edge of the red rays in the rainbow band of colors resulting from sunlight that had passed through a glass prism. Since a rise in temperature was recorded, Herschel correctly inferred the existence of radiation in the infrared. Two years later, Wollaston, natural scientist extraordinary—astronomer, chemist, physicist, geologist, and metallurgist—proved the existence of ultraviolet rays. He based his conclusion on an experiment in rudimentary photography—the darkening of silver chloride in the presence of radiation. Clerk Maxwell described the character of radio waves some 30 years before Hertz produced them in the laboratory.

☉

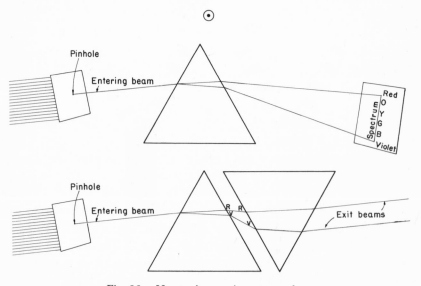

Fig. 26.   Newton's experiment on colors.

The character of light was superbly demonstrated many years ago, by no less a person than Sir Isaac Newton, who in 1672 proved to his own satisfaction, if not to that of his contemporaries, that color was an intrinsic property of radiation. Previously, scientists had generally supposed that the colors of the spectrum were properties conferred by the glass of the prism. Newton's experiment was simple. He showed that, where a single prism would break up sunlight into its color spectrum, a second prism, set with its apex opposite to that of the first, would cause the colors to exit in parallel directions and essentially reunite (Fig. 26), to reproduce a beam of white light. Thus the sensation of "white" is really due to the presence of all colors.

Theories of Light and Radiation

From the time of Newton until the beginning of the nineteenth century, progress in optics and in the understanding of radiation was slow. There were some experiments and much speculation concerning the nature of radiation. Two theories held the field. One claimed that light consisted of myriads of tiny particles or corpuscles. The sizes of the individual bullets, or perhaps their relative masses, were supposed to determine the particular color sensation recorded by the eye. The second theory asserted that light was a wave motion, with the colors fixed by the spacing between crests of successive waves.

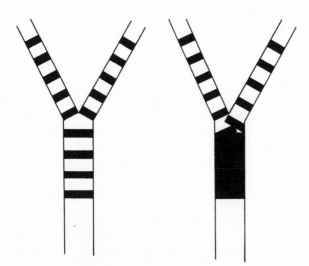

Fig. 27.   Interference of waves entering a canal.

Although Newton actively supported the corpuscular theory, his writings show that he never completely abandoned the wave concept. He apparently conceived of vibrations either accompanying the particle in its flight or set up by it on entrance into some material medium. And, in thus straddling the fence, Newton was—as I shall indicate presently—almost 250 years ahead of his time.

Early in the nineteenth century, Young made the first decisive experiment, which seemed to establish beyond question the wave character of radiation. He discovered that two identical beams of light, when superposed, would exactly cancel each other under certain special conditions (Fig. 27). Light plus light equals darkness! Indeed a strange equation, and apparently quite irreconcilable with a corpuscular theory! But the wave theory offered some hope. For the downward motion of the waves in one beam might exactly neutralize the upward motion of those in the other.

A simple analogy may make this point clearer. Two men trying to fell a tree with a two-handled crosscut saw will struggle in vain if they pull simultaneously and push simultaneously. The saw will remain stationary. The two light beams are like the two men. The light waves push and pull at the light-sensitive surface of the eye. But, because the motions cancel, as in the analogy, the net result is no motion whatever—and a final sensation of darkness.

Many physicists extended the observations of *light interference,* as the phenomenon came to be called. The wave theory seemed to explain almost all of the known properties of radiation. The spacing between wave crests—the wavelength—determined the color. Measures showed that a beam of red light contains about 40,000 waves per inch, whereas a beam of violet light contains about 70,000, almost twice as many. All of the rainbow colors fall between these two values. Clearly, then, ultraviolet radiation had even a smaller wavelength than the violet, and infrared longer than the red.

Theoretical studies by Clerk Maxwell, about 1860, suggested that light waves were electromagnetic in character. Scientists knew that an electric current, flowing through a wire, produces an extensive magnetic field in the surrounding space. When the current is cut off, turned on, or reversed, a sharp magnetic pulse emanates from the wire and spreads out into space, with a speed, as Maxwell proved, equal to that of light. Maxwell further suggested that if the current could be reversed back and forth rapidly enough, in a wire about as long as a light wave, the fluctuating electromagnetic field outside the wire would be identical with the phenomenon of light.

Now, light travels with a speed of 186,000 miles, or about 10^{10} inches, per second. With 40,000 waves per inch, we find that there are some 4×10^{14} individual wave crests in a 1-second flash of red light, each crest corresponding to a complete electrical cycle in the imagined wire. I do not wish to stress the size of the number unduly except to point out that so high a frequency was, until recently, beyond the power of man to produce or control by a method of oscillation. But W. W. Salisbury and E. M. Purcell recently demonstrated a method for making electrons oscillate at such frequencies. And, as expected, light of the appropriate color escaped from the apparatus.

If, instead of 4×10^{14}, the number of cycles were as small as 1 million per second, the waves, so produced, would be correspondingly longer than those of light, about 1000 feet long, an infra-infrared radiation. Analogously, the radiating wire for carrying the pulsing electric current should be about the same length. The reader may already have surmised that I am now discussing a well-known type of radiation: radio waves. Maxwell predicted their existence long in advance of discovery, by a train of logic not unlike the foregoing.

In 1888, nine years after the death of Maxwell, Hertz successfully produced radio waves in the laboratory. Thus did the imagination of a theoretical physicist and the ingenuity of a skilled experimen-

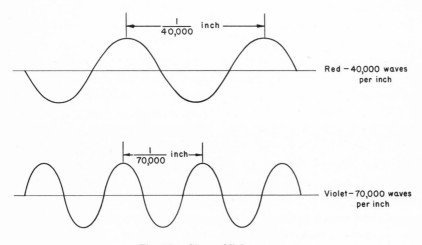

Fig. 28.   Sizes of light waves.

talist create the beginning of an enormous industry. Further investigations, especially those connected with the wartime development of radar, closed the gap between the long radio waves and the much shorter heat waves of the infrared. At the other end of the spectrum, scientists pushed the boundaries to the x-rays, so minute that they can traverse the spaces between atoms in seemingly solid matter and disclose in shadow pictures the structure of the inside of an object—a diseased bone or defective iron casting.

All of the radiations, as Maxwell had forecast, proved to be electromagnetic, a form of light. Our eyes are sensitive to only a tiny fraction of the entire band. They are blind to x-radiation and to the infrared. And our ears are deaf to all electromagnetic waves, including the radio waves—which is a blessing; otherwise we should never be free from the multitude of radio and television programs that clutter the ether.

But, if radio waves are produced by electric currents oscillating back and forth in a wire, how are the light waves formed? We have seen that the antenna must be of atomic dimensions for light. And therein lies the answer! The atom itself, although actually even smaller than the waves it produces, is really a miniature transmitting station. All matter is electrical in nature. Each atom contains electrons, which oscillate back and forth, "broadcasting" light waves.

The analogy between atoms and radio transmitters is so exact that I venture to push it still further. Each commercial transmitting station has its own assigned frequency or wavelength, by which we may identify it. When you tune the receiver to a definite setting, you usually know in advance what station you will hear. Similarly, each kind of atom has its own characteristic wavelength. By measuring the wavelength, and thereby determining the exact color of the light, we may identify the transmitting atom, and decide whether it is hydrogen, oxygen, sodium, iron, or what.

Unlike the radio station, however, an atom may send out not one

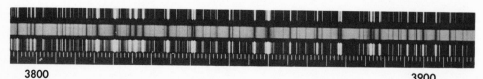

3800   3900

Fig. 29.   Spectrum of iron between 3800 and 3900 A: (*center*) absorption lines from an exploded wire; (*top and bottom*) comparison emission spectrum from an iron arc. (Mount Wilson Observatory.)

but a hundred or in some elements many thousand characteristic wavelengths. Furthermore, some of the frequencies are more "popular" than others, so that many atoms may radiate simultaneously on one wavelength with only a few on another. Consequently, the various colors emitted by a group of atoms may differ markedly in intensity.

The chemist, in his laboratory, puts these properties of atoms to effective use. He has, let us say, a sample of steel, which he suspects may contain zinc. This impurity may impair the strength of the finished product, even if it is present only in minute quantities. No chemical test is sensitive enough to find the zinc. He would be balked without the aid of light. But, now, the chemist chips a tiny sample, vaporizes it in an electric arc or spark, and carefully studies the light that comes from the incandescent material. Zinc, if present, would betray itself by the characteristic wavelengths of its radiations, which are entirely different from those of iron (Fig. 29). In a similar way, as we shall see presently, the astronomer may examine light from the sun and other heavenly bodies and determine what chemical elements compose them.

⊙

The instrument that one uses for such analysis is called a *spectroscope* or *spectrograph,* according to the method of observation—visual or photographic. The principle of the device is simplicity itself. Suppose that we wish to analyze the light from a luminous flame. We place, first of all, a narrow slit, *S,* in front of the flame (Fig. 30).

The Spectrograph

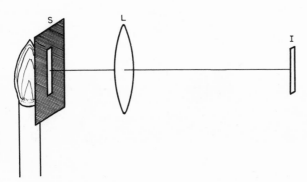

Fig. 30.   A lens produces a single image of a slit.

A simple camera lens, *L*, forms an image of the slit at *I*, but all the colors come to the same focus.

Now, in front of the lens place a small glass prism, *P* (Fig. 31). The colors bend at different angles, so that the red comes to a focus at *R*, the violet at *V*, with the complete series of rainbow shades in between. If the flame is emitting two distinct colors, say one in the yellow and another in the red, so that it looks orange to the eye, the spectroscope will break up the colors, forming one slit image at *R* and another at *Y*. We have one slit image for each color sent out by the flame. Because the individual images look like fine lines, physicists customarily call them the "lines" of the spectrum. I should mention that a somewhat better arrangement, optically, employs two lenses instead of one, as shown in Fig. 32.

Green light has a wavelength of about 0.00005000 cm, or about 0.00002 inch. The red line of hydrogen has a wavelength of 0.0000656279 cm. The centimeter is thus not a very satisfactory unit for expressing measurements of light waves. We require too many zeros in front of the significant numbers. To avoid this difficulty, scientists have agreed to use the angstrom unit (A) as the basic length. This unit is named after A. J. Angstrom, a Swedish physicist who pioneered in spectroscopic research; it equals 0.00000001 cm or 10^{-8} cm. One hundred million of these angstrom units are required to span a single centimeter.

In our new unit, the red hydrogen line, known as H-alpha, measures 6562.79 A. X-radiations, on the same scale, may range

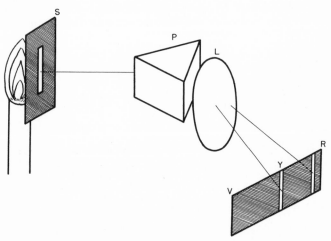

Fig. 31.   A prism produces a spectrum of multiple slit images.

from 50 to 100 angstrom units in length down to waves as small as 0.1 A. This value by no means sets a short-wave limit to radiation. The so-called gamma rays emitted by radium, uranium, and other radioactive atoms overlap with the shorter x-rays and extend to 0.03 A. Even shorter wavelengths are produced in atomic transmutations, including those induced by cosmic radiation.

The wavelength, as measured, depends upon the velocity as well as upon the character of the emitting atom. The atom, sending out successive waves at regular intervals, may be likened to an ardent lover who has agreed to write his beloved at the rate of one letter every hour. The vagaries of mail service impair the analogy, but we shall assume that he has a retinue of couriers waiting to transport each finished epistle. The messengers travel at constant and identical speed, and travel directly to the destination.

Now, as long as the addressee receives the letters at the rate of one per hour, she is assured of the constancy of her lover. Suddenly the interval lengthens to one hour and ten minutes. Does this increase denote a cooling of the affections? Not at all. Her swain is merely traveling away from her. The letters, themselves still dispatched one per hour, take longer to arrive. When he reaches his destination the letters again arrive at the normal rate. But when he starts on the return journey the interval is less than one hour.

The rate of arrival measures the actual distance between the letters in transit. The distance changes with the velocity of the letter writer. In precisely the same way, an atom that is moving away from

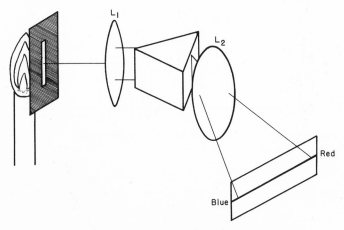

Fig. 32.   Schematic diagram of a prism spectroscope.

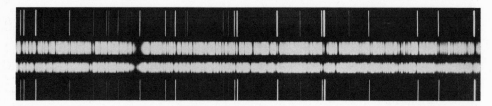

Fig. 33.   Two spectra of the constant-velocity star Arcturus, taken about 6 months apart. The velocity difference, shown by the Doppler shift of the spectrum lines, amounts to 50 km/sec, and is entirely due to the orbital velocity of the earth. (Cf. p. 18 and Fig. 12.) (Mount Wilson Observatory.)

the observer causes the arriving waves to lie farther apart. The wavelength is longer. An atom that is approaching the observer will produce a shortened wavelength.

The astronomical spectrograph measures the wavelength as it arrives. If a given line lies on the red side of its normal position, the emitting atoms are receding from the observer; if the line lies on the violet side, the atoms are approaching. The magnitude of the shift reveals the speed. Suppose that H-alpha, normal wavelength 6562.79, were shifted to the red by 1/1000 of its wavelength, or 6.56 A. The resulting value of 6569.35 A. would indicate a recessional velocity of 1/1000 the speed of light. The hydrogen atoms, therefore, would be moving away at 186 miles per second.

The principle has many applications in astronomy. One derives the speed of stars, nebulae, or various portions of the solar disk from such measurements of spectrograms. We shall have many occasions to refer to this shift in wavelength of spectral lines, termed the *Doppler effect*. (C. J. Doppler, an Austrian physicist, was the first person to predict the existence of the shift in color now called by his name.)

⊙

Modern Theories of Light

Almost all of the foregoing discussion relates to the wave properties of radiation. The physicists of the nineteenth century seemed to have completely killed the corpuscular theory of light. No one dreamed that its ghost was due to rise again and haunt the halls of twentieth-century laboratories. In truth, the corpuscular theory was far from dead. And in the interval from 1900 to 1925 its ancient skeleton, revitalized with the solid flesh of experimental science, almost completely deposed the wave theory from its position of high favor.

Max Planck, at the turn of the century, reopened the question. His investigations, together with the later studies of Niels Bohr, a Danish physicist, reestablished the view that light behaved like miniature bullets. An individual light projectile they called a *quantum,* and from all sides came experimental evidence supporting the quantum theory.

We may gain some insight into the reasons behind this development if we refer once again to the radio-transmitter analogy. Your ear immediately distinguishes between two broadcast stations of equal power that lie at different distances. The farther one is (usually) fainter. The waves reaching your receiver are merely less intense. Otherwise they possess all of the qualities of the waves from the nearer station.

But if radio energy were like a blast of bullets from a machine gun, shot from the transmitting antenna, a very different effect would result. The projectiles from a nearby station might arrive so nearly continuously that your program would come in unimpaired. Bullets would always be present in your neighborhood. But, from the very distant station, only an occasional bullet would fall within the range of your receiver. At the moment, you would hear a word or two as clearly as from the neighboring transmitter. Between bullets, your radio would be completely silent. Analogous experiments, performed with beams of light, clearly showed the quantum, or bullet, character of the light. Nevertheless, the older experiments, indicating just as plainly the wave nature of radiation, remained unchallenged. Here was indeed an impasse! How could light behave like a projectile in one instance and like a wave in another?

About 1925, de Broglie and Schrödinger hit upon a compromise view. In solving the problem, they proposed a second paradox, just as puzzling as the first. De Broglie showed that matter itself, hitherto considered to be entirely corpuscular, also could display wave characteristics. The waves simply guided the particles—light quanta or electrons. Newton himself had forecast the answer, as I have previously indicated, by suggesting that some form of waves controlled the corpuscles in their flight.

⊙

Atomic Structure   Our concept of atoms has undergone enormous revision with the passing centuries. The ancients visualized atoms as tiny geometric figures: spheres, cubes, pyramids, and so on. The possibility of their possessing an internal structure was not even considered because such a structure would imply smaller units, whereas the term "atom" was derived from a Greek word signifying "indivisible."

For the early chemist, concerned mainly with explaining the *differences* between chemical elements—for example, why iron is unlike copper—the simple geometric forms were quite sufficient. But, when Mendeleev, in 1871, emphasized the chemical *similarities* between certain elements, the idea of atomic structure was born. For how could similarities exist in bodies that possess nothing in common?

To illustrate the problem that confronts us, I list the 18 lightest of the more than 100 known atomic varieties, in the order of their weights. Hydrogen, the least massive, comes first, helium second, and so on:

Hydrogen							Helium
Lithium	Beryllium	Boron	Carbon	Nitrogen	Oxygen	Fluorine	Neon
Sodium	Magnesium	Aluminum	Silicon	Phosphorus	Sulfur	Chlorine	Argon

I have arranged the elements in this tabular fashion, leaving the long semiartificial gap between hydrogen and helium, because each column now contains elements that resemble one another chemically, and, sometimes, physically as well. Lithium and sodium are both light metals. Fluorine and chlorine are odorous, highly corrosive gases. Helium, neon, and argon, all gases, enter into no chemical unions. They are the misogynists of the atom family.

The importance of this tabulation will become much clearer if we consider the stable combinations that various elements form with hydrogen. Hydrogen combines with itself to produce a molecule H–H or H_2; similarly, the atoms of the same column combine with hydrogen to give LiH, and NaH, lithium (Li) and sodium (Na) hydrides, respectively. Beryllium and magnesium, on the other hand, each require two atoms of hydrogen in order to attain the full satisfaction of chemical union. The hydride formulae are BeH_2 and MgH_2. Atoms of the next column require three hydrogen atoms apiece, BH_3 and AlH_3, while those of the fourth column form the compounds CH_4 (marsh gas) and SiH_4.

One might now expect the hydride of nitrogen to have the formula NH_5, and, indeed, nitrogen and phosphorus often possess a chemical combining power of five. Usually, we find a decrease and obtain NH_3 (ammonia) and PH_3 (phosphine). The hydride of oxygen, OH_2, is more familiar to you as H_2O (water). And it may come to you as something of a shock to note the kinship between common water and hydrogen sulfide, H_2S, the gas with a rotten-egg odor. The atoms of column seven give the hydrides HF (hydro-fluoric acid, sometimes forming the compound molecule H_2F_2) and HCl (hydrochloric acid). Thus, dropping from one to zero for the final column, we readily see that the failure of this group of atoms to combine chemically is in keeping with its position in the tables. Figure 34 displays, in graphic form, the molecular relations involved.

The foregoing discussion of chemical properties has, perhaps, verged on the technical. But it is important to see that atomic structure must account for many diverse properties of matter. I might have traced the similarities through the entire table of elements. To exhibit the full relation I should have required not merely two or eight columns, as shown above, but 18 and 32 as well, and I shall not attempt to detail here how the heavier elements fit into the picture.

The chemist likes to visualize his atoms as having hooks or handles to which other atoms may attach themselves and thus form molecules. We may compare hydrogen, lithium, and sodium, to small, medium, and large one-handled cups. Beryllium and magnesium are varieties of two-handled sugar bowls. Carbon, a four-handled vessel, forms a satisfied chemical compound with hydrogen when one "cup" is attached to each of the handles.

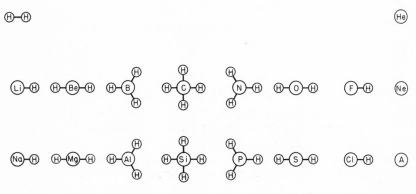

Fig. 34.   Schematic form of molecules.

So much for the chemist's viewpoint. The physicist is more concerned with the atom's ability to absorb and emit various kinds of radiation. He wants to poke inside and find out what is going on. Unfortunately, the atom is so small that we cannot pry into its secrets by ordinary methods. Even light waves are too gross for the examination. Trying to "see" an atom is a little like trying to use a stove poker for a toothpick.

Nevertheless, strong-arm methods have revealed a great deal. We may blast the atoms apart and see what comes out of the wreckage. Sir J. J. Thomson did this very thing, as long ago as 1897, and found the electron, a minute particle of negative electricity, weighing only about 1/2000 as much as a hydrogen atom. He discovered, further, that the electrons from different kinds of matter were identical. Here, indeed, was progress! Electrons were one of nature's fundamental building blocks.

The second great contribution also came from the Cavendish

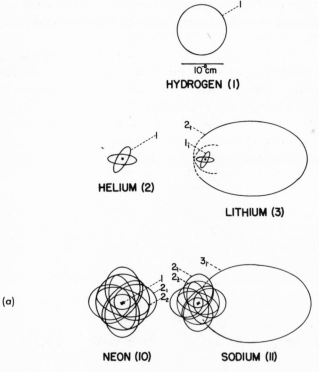

Fig. 35. (*a*) Orbit models for several lighter atoms. (*b*) Orbit model for the

Laboratory, at Cambridge, England, where Sir Ernest Rutherford proved that the massive and positively charged part of the atom occupied a minute volume at the center, like a schoolboy's marble in the middle of a large dance hall. The electrons of the atom, mere gnats by comparison, were flying somewhere in the vast volume of the hall. The atom is mostly empty space and the apparent solidity of matter an illusion produced by the grossness of our senses.

Only gradually has the rest of the atomic picture emerged. The chemical nature of the atom depends ultimately upon the number of positive-charge units in the nucleus. Each nucleus attracts to itself a number of negative electrons just sufficient to neutralize its positive charge. Thus hydrogen possesses one external electron, helium two, and so on throughout the table of elements.

Radium, element number 88, has 88 electrons (Fig. 35*b*). But how are the electrons arranged and why do they not collapse entirely upon the nucleus? For some years physicists supposed that the electrons were revolving around the central nucleus: the atom, a solar system in miniature. Many orbits were possible and radiation was supposed to be emitted when an electron fell from an outer orbit to one nearer the nucleus.

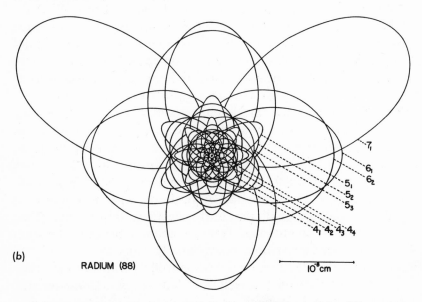

(b)   RADIUM (88) 10^{-8}cm

radium atom; the numbers symbolize the sizes and shapes of the different interlocking orbits of the 88 electrons. (Courtesy Wendt and Smith, *Matter and Energy*, The Blakiston Company.)

But there still were many difficulties. How could light emitted in this erratic fashion possibly possess any wavelike property? There was nothing suggestive of the radio transmitter in this model. Furthermore, it did not satisfy the chemists, who thought they needed something steadier than a swiftly moving electron to act as an atomic handle.

Schrödinger and his many followers put the wave into the atom. The atom, like a musical instrument, is full of vibrations. I cannot tell you what is vibrating, however. Some physicists aver that the electron itself dissolves into a pulsing electric mist, thus doing away entirely with solid charged particles. Other scientists believe that the electron retains its particle identity and that the vibrations are "probability waves." This sounds highly complicated, but the idea is certainly not unfamiliar to you. We often speak of a crime wave or a suicide wave, in much the same sense. In the midst of such a wave the probability of any one person's committing a crime or taking his life is somewhat greater than under normal conditions. Similarly, within the atom, the electron is more likely to be found where the probability wave has its greatest intensity.

In Fig. 36, we see a number of such probability waves photographed from a model hydrogen atom. We do not know where the single electron is. Indeed, the new theory provides the interesting corollary that we can never locate the particle exactly. Our attempts to pry into the private life of an electron cause it to deviate from normal behavior. We only know that, on the average, the electron will lie more often in the regions of denser probability than elsewhere.

Note that numerous vibrational states exist, one for each of the pictures. Like the string of a violin, the atom may resonate in an infinite number of ways. Unlike the musical instrument, however, which radiates continuously, the atom transmits its vibrations to external space in the form of light only during the changes from one pulsating pattern to another. And, during these transitions, we can trace an actual oscillation of the electron, akin to that of the electrons in a radio antenna. Thus the analogy between the atom and a radio transmitter becomes essentially complete.

The atoms prove to be receivers, as well as transmitters. They are capable of absorbing as well as radiating light. Consequently, if we pass white light through a gas, the atoms will absorb their own characteristic radiations. The spectrum of the white light will show gaps—dark lines—where these energies are missing. We shall see

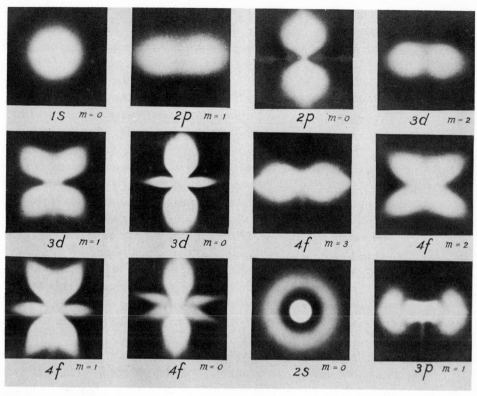

Fig. 36.   The hydrogen atom wave model. The numbers and letters are the physicist's notation for the different vibration patterns. (H. E. White.)

presently that the sun's atmosphere acts in precisely this fashion to
produce an absorption spectrum—a dark-line counterpart of the
bright-line spectrum that we should see if the background source of
white light were removed.

The wave picture of the atom is acceptable to the chemist as well
as to the physicist. The two electrons of helium, for example, form
a perfect sphere with their interwoven pair of probability waves.
The external smoothness of the pattern gives no hook or handle to
which another atom may cling. Hence the chemical inertness of
helium.

Hydrogen, with only one electron, lacks the symmetry of helium.
And it seems to yearn for this perfection. I do not wish to imply
that the atom has a mind of its own. A drop of rain may be said
to "yearn for the sea" in much the same impersonal fashion. The

atom, like the raindrop, attains greater stability when its "yearning" is satisfied. Two hydrogen atoms, sharing one another's electrons, achieve the sought-for symmetry and a stable hydrogen molecule results.

Lithium, atom number three in the table, has three electrons to arrange. Two will fit, as for helium, in a sphere around the nucleus. If I may again shift the metaphor, the ringside seats are now filled and the third electron is assigned, along with its probability wave, to the first balcony, where it wanders lonely and unsatisfied. A hydrogen atom, coming into the vicinity, may seize upon this external electron, which lithium gladly shares because the remaining portion possesses a perfect form. The atom pair, thus bound together, forms a molecule of lithium hydride.

Although only two seats are available at the ringside, there are eight in the first balcony. Thus the spherical form of eight interlocking probability waves does not appear before atom number 10, inert neon. The enormous chemical activity of fluorine, atom number nine, arises from its frantic search for an atom having one spare electron to fill the gap. Fluorine will even dissolve glass in its effort to attain the symmetrical form. Oxygen, needing two electrons, is almost but not quite so active chemically as fluorine. Its common combining power (or valence, in the language of the chemist) of two is clearly in keeping with the number of missing electrons, to be furnished by two hydrogen or other convenient atoms.

I shall follow this argument just one step further. We must place the eleventh electron of sodium in the second balcony, because the other ten electrons occupy all the available nearer seats—two at the ringside, eight in the first balcony. Now, notice the physical resemblance between sodium and lithium, the former with a lone electron in the second balcony and the latter with one in the first balcony. No wonder that the two substances behave in analogous chemical fashion!

A normal atom is electrically neutral. Take magnesium, for example, whose dozen negative electrons exactly cancel the twelve units of positive charge in the nucleus. Atoms in a gas are constantly colliding with one another, and under certain conditions, for example in the hot atmosphere of the sun, where the atoms are moving very fast, the violence of the interatomic collisions may tear away one or more of the outer electrons. Magnesium with only eleven, instead of its customary twelve, outer electrons resembles sodium. The magnesium atom, however, is no longer neutral but possesses a net positive charge. Hence we term it an *ion*. We designate the neutral atom of magnesium by the symbol MgI, the singly ionized atom by MgII, the doubly ionized atom by MgIII, and so on.

The spectrum of singly ionized magnesium bears some resemblance to that of sodium, the difference arising from the fact that the atomic nuclei are not identical; magnesium is heavier. The radiations of doubly ionized magnesium are, analogously, akin to those of neon, and so on. Finally, magnesium with eleven electrons gone and one remaining radiates exactly like hydrogen, except that all the spectral lines lie far to the ultraviolet—in the x-ray region—with frequencies 12×12 or 144 times greater than those of hydrogen.

Some atoms, like sodium and calcium, lose one of their outer electrons quite easily. Hydrogen and oxygen are harder to ionize, with helium the most difficult of all. Whenever we see, in the spectrum of any heavenly body, the radiations of ionized helium, we infer immediately that the atoms are being subjected to enormous excitation. The temperature must be extremely high, probably in excess of $20,000°K$ ($45,000°F$). At lower temperatures, the moving atoms would not collide with sufficient energy to tear the electron away. (The letter K indicates the Kelvin scale, a Centigrade scale with the zero set at $-273°C$; F refers to the Fahrenheit scale. $0°C = 273°K = 32°F$, the freezing point of water. $100°C = 373°K = 212°F$, the boiling point of water at sea level.)

A molecule, like an atom, can radiate, but in a more complex manner. Not only can the wave patterns of the electrons change, but the atomic nuclei, vibrating back and forth like balls connected with a piece of elastic or revolving around each other in orbits, radiate whenever they change their vibrational or rotational states. Indeed, they may shift simultaneously in all three patterns. Where an atom will produce a single line, a molecule will give many, often so close together as to give the appearance of a band (Fig. 37). Each molecule has its own characteristic spectrum, from which we may identify its presence. And, like atoms, molecules may, under favorable circumstances, also exist in the ionized condition.

The philosopher may see in the chemical behavior of matter some urge to self-completion; the poet, a personalized force governing atomic "mating." But the polygamous character of certain elements somewhat mars the romatic picture; and the great frequency of "divorces," the violent disruption of molecules in disastrous collisions, introduces a further discordant note.

The chemist, however, sees the hurly-burly in his test tube as a conflict of forces rather than of personalities. The laws of chemistry are as inescapable as the law of gravitation. A rock falls down a mountainside. One atom captures another and forms a molecule. Both events are expressions of the same fundamental principle: a

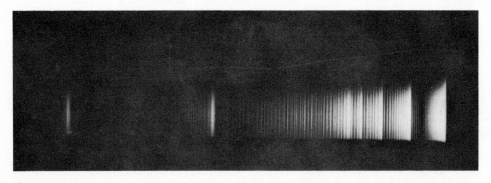

Fig. 37.   The spectrum of molecular cyanogen. These bands of CN are strong in many stellar spectra, including that of the sun. (Harvard Physics Laboratory.)

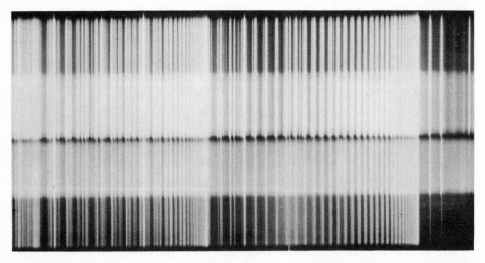

Fig. 38.   Banded (fluted) spectrum of molecular carbon superimposed on the line spectrum of the element ruthenium. (Massachusetts Institute of Technology.)

tendency of unstable objects to move into positions of greater stability. The laws, if properly formulated, are presumably universal. The atoms are the same wherever we find them—on earth, in the sun, or in other great galaxies like our own Milky Way system, millions of light-years distant.

The scientist's faith in the universality of nature's laws gives him confidence to attack the problems of remote heavenly bodies. The stars send us light. Analysis of that light tells us what atoms are present and the nature of their physical surroundings. More than that! Astronomical observation contributes fundamentally to our knowledge of the nature of matter and radiation. The sun and stars serve as laboratories, where matter may exist under conditions very different from those we can produce in our terrestrial laboratory. In the depths of the universe atomic encounters can be far more violent than any we find on earth—even in the heart of a hydrogen bomb.

Tennyson, in his "Lucretius," vividly portrays this cosmic conflict, this dynamic essence of matter:

> Storm, and what dreams, ye holy gods, what dreams!
> For thrice I waken'd after dreams. Perchance
> We do but recollect the dreams that come
> Just ere the waking: terrible! for it seem'd
> A void was made in Nature; all her bonds
> Crack'd; and I saw the flaring atom-streams
> And torrents of her myriad universe
> Ruining along the illimitable inane,
> Fly on to clash together again, and make
> Another and another frame of things —

4

The Message of Sunlight

Probably the simplest and best-known fact about the sun is that it sends us light and heat. Deep in the sun, near its center, some sort of process generates energy, which, relayed from atom to atom, eventually worms its way to the solar surface. Then, suddenly finding itself free, the ray of energy speeds across the 93 million miles of space, plunges through the atmosphere, and strikes the earth's surface. Only one part in 2000 million of solar radiation actually reaches the earth. The remainder, except for the minute fraction intercepted by other planets, is lost forever to the solar system. The rays speed on out into the realms of interstellar space, to which their ever-dwindling intensity imparts scant warmth.

The radiation that falls on the earth's surface carries many cryptic messages. The astronomer has the task of decoding them, of seeking out what they mean. The key to the cipher, as we shall see, lies in the laws of physics. In this chapter we shall be concerned with the deciphering of only a small portion of the full message. In each of the subsequent chapters of this book we shall deal with a different aspect of the information derived from sunlight.

We all are conscious of the two most obvious facts of sunlight: its quantity and its quality. The quantity varies with the season. Clouds, dust, or smoke in the atmosphere may sometimes lessen the intensity of the radiation reaching us. But, if our observations are to mean anything in terms of the sun, we must find some way of eliminating or correcting for these local conditions.

I have already mentioned that sunlight is "white," which gives a picture—though a very crude one—of its quality. "White" is too indefinite a term, covering a multitude of shades, as anyone who has tried to match a sample of white cloth will substantiate. We have had to devise more exact definitions of both quantity and quality, and to invent some experimental way of measuring them precisely.

⊙

Brightness of the Sun

Let us consider first the question of intensity. How should we express our measures? Numerous possibilities present themselves, many of them, unfortunately, far too technical for presentation here. But there is one measure of brightness that should be familiar to everybody: candle power. Some years ago this term was used even more generally than at present, to indicate the brilliance of light bulbs. A 30-candle-power lamp was expected to produce as much light as 30 "standard candles." Of course we must have some standard for the candle, specifying its size, its composition, the length of its wick, and even the amount of air it is to consume each minute while it burns.

We now measure the power of light bulbs in terms of the electric energy they consume. But "watts," in themselves, are not a good measure of the resulting illumination, because, in the ordinary filament-type lamp, a large percentage of the electric energy goes into infrared, or heat radiation, an utter waste as far as light is concerned, because it is invisible. The familiar Mazda lamp gives approximately 1 candle power per watt, almost four times as much light for the same energy as the old inefficient carbon-filament variety yielded.

Neon and other advertizing signs, consisting of tubes filled with gas made luminous by an electric discharge, give a still higher percentage of visible light per watt. Since a much greater fraction of the electric energy appears as light and very little as invisible radiations, a 15-watt lamp gives as much candle power as a 60-watt tungsten-filament lamp.

Now, which of these designations, candle power or watts, are we to use in describing solar radiation? Clearly they are not synonymous. If we wish to refer to *visible* light, we obviously must use candle power, which defines what we actually see. But, as I pointed out earlier, much of the solar energy lies in the infrared and ultraviolet regions—energy completely left out when we use our half-blind eyes to measure the intensity of sunlight. Now, the 60-watt bulb is actually radiating 60 watts of energy, even if only 2 percent produces an effect on the eye. Hence the watt is really a very convenient unit for expressing the total rate of radiation in all wavelength regions.

Let us consider how we might go about measuring the sun's candle power. First of all, in a dark room, we might set up a standard candle and judge the intensity of light falling from it upon a piece of white paper 1 foot distant. Next, we blow out the candle and admit sunlight into the room, adjusting its intensity on the paper with filters of dark glass until it exactly matches that from the candle. We may calibrate these filters with the aid of photocells.

If we find it difficult to remember the intensity, we may set up the candle on one side of the paper and allow sunlight to fall on the other. Put a tiny spot of grease at the center of the sheet. This spot causes the paper to become semitransparent, so that some of the sunlight comes through. We now compare the brightness of this spot with that of the rest of the sheet. If the spot appears bright or dark we know that the light sources are unequal. Hence we adjust the intensity of the sunlight until the spot seems to disappear (Fig. 39).

To bring about this equality of illumination, we shall find that our dark-glass filters will have reduced the intensity of sunlight by a factor of some 10,000 times. The sun at the zenith, therefore, produces an intensity of about 10,000 foot candles, that is, 10,000 times more luminous energy than that from a candle 1 foot distant. The sun is 5×10^{11} feet (500,000 million feet) distant. Hence, taking into account that the intensity of radiation varies inversely as the *square* of the distance, we readily figure the total candle power of the sun to be about 2.5×10^{27}.

This figure is inconceivably great. When we correct for the dimming effect of the earth's atmosphere, the value comes out even larger, about 3×10^{27} candles. To manufacture this number of candles, even of tiny birthday-cake variety, we should require a lump of tallow 10 times larger than the earth. The wicks alone would span the surveyed universe of some billion light years about ten times. (A light year is about six million million (6×10^{12}) miles—the distance traversed by light, at the speed of 186,000 miles a second, in 1 year.) The candles, crowded to the touching point, would cover a birthday cake as big in diameter as the orbit of the earth.

Each square inch of the solar surface is shining with the intensity of 300,000 candles. No wonder that the sun hurts our eyes when we look at it directly!

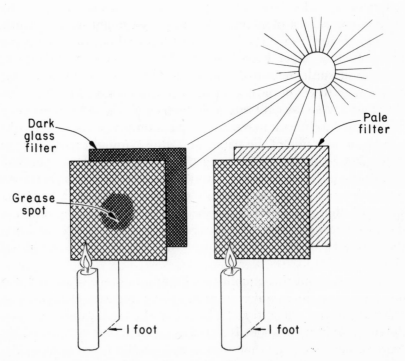

Fig. 39.   Schematic experiment for measure of the sun's candlepower. Sunlight, reduced by a dark filter, gives illumination equal to that from a standard candle.

The experimental procedure I have just described for measuring the brightness of the sun is very crude. A better device for registering the intensity of radiation would be a photographer's exposure meter, or, even better, one of the illumination meters that the electric light company will bring into your home to see how many lamps you should have. These recorders generally respond to a much wider color range than does the human eye. We therefore obtain a better measure of the visible brightness of the sun if we put a piece of light-green cellophane over the recorder, since the eye is more sensitive in the green than elsewhere.

⊙

The Solar Constant

A great sensitivity range, however, is precisely what we require if we are to measure the total energy of the sun, rather than just the fraction we perceive as light. The photocell is better than the eye, but the best instrument of all is the *thermocouple*. It consists merely of two wires—a "couple"—of dissimilar metals, iron and copper, for example, or bismuth and antimony, joined together. The opposite ends are connected to an ammeter. Now, when heat is applied to the junction, an electric current results. The amount of this current, registered on the ammeter, gauges the quantity of heat.

Here we have a possible measuring instrument. We affix to the thermocouple a small metal disk to act as a receiver for the radiation. Next we carefully coat the surface of the disk with a layer of black, because pure black absorbs all of the colors that fall upon it, infrared, visible, and ultraviolet alike. The energy absorbed heats the disk and the junction; the amount of current measures the radiation falling upon the receiver. Astronomers have built compound thermocouples of very high precision and great sensitivity (Fig. 40). With one of these devices attached to a large reflecting telescope, one could detect the radiation from a tallow candle 100 miles away. And the energy from a hot spring a few hundred feet across on the moon—if such existed—would cause a large deflection of the ammeter.

Now, with the thermocouple instead of the eye as a measuring instrument, we repeat our earlier experiment, after replacing the candle by a lamp of known wattage. The measurements lead to the following interesting data concerning the sun's total radiation. The

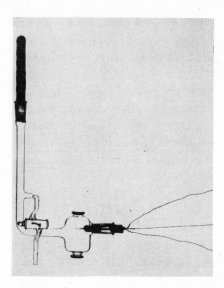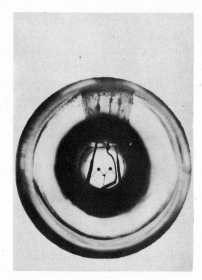

Fig. 40.   Delicate thermocouple for measurement of stellar and planetary radiation. The thermocouple has two junctions, each capped with a round black
absorbing surface. If the image of a star is moved from one junction to the other,
the current reverses, thus doubling the galvanometer deflection that one would
obtain from a single junction. (Mount Wilson Observatory.)

energy, outside the earth's atmosphere, is 1100 watts, or 1.1 kilowatts, per square yard. Therefore, we calculate that the total energy
production of the sun amounts to 3.86×10^{26} watts, or 40,000 watts
per square inch of its surface.

Concealed in the foregoing figures are some very important facts,
which I shall now point out. The sun, we figure, gives us 300,000/
40,000 or about 8 candle power for every watt. A 100-watt Mazda
lamp produces about 1 candle power per watt, whereas the old
carbon-filament lamp required almost 4 watts for the same illumination. A candle expends about 8 watts of energy to produce its
single candle power. The question is: wherein lies this difference in
relative efficiency? Why is the sun more than 60 times as efficient
as the candle in giving off luminous radiation?

The wattage tells us the total energy radiated. The candle power
indicates what portion produces the sensation of light. Most of a
candle's radiation lies far in the infrared where the eye cannot see
it. The temperature of the candle flame is only 1930°K (2990°F);
that of the carbon filament, about 2115°K (3300°F); that of the

tungsten filament, 2760°K (4510°F). Theory shows—and experiment bears out its conclusions—that the hotter the source of light, the greater the proportion of its radiations that will be emitted in the shorter wavelengths.

In Fig. 41, for example, we see plotted the distribution of intensity with wavelength of 1 watt of energy emitted by sources of selected temperatures. The curves differ principally in that the maxima for the cooler temperatures lie toward the longer wavelengths. The observed shape of the curve for the sun agrees closely with the theoretical curve for a body with a temperature of 6000°K. Hence we infer that the temperature of the sun is 6000°.

The shaded area of Fig. 41, with the sharp·peak, indicates the portion of the solar energy perceived by the human eye. Only 14 percent is thus recorded; small as this amount is, it is nevertheless far greater than the fraction perceived from the lower-temperature sources. The diagram makes clear why the tungsten filament is so much more efficient than one of carbon. The latter would be equally bright, if we could heat it to 2760°K without vaporizing the filament.

One further item of interest is that a 2-watt fluorescent lamp,

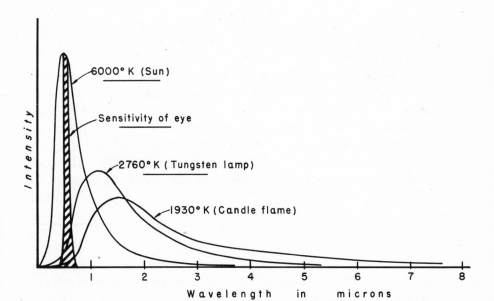

Fig. 41.   Distribution of color for sources of various temperatures. The shaded area indicates the sensitivity of the normal eye. (1 micron = 10,000 angstroms.)

radiating only in the green where the eye has its greatest sensitivity, would be as bright as an ordinary 100-watt bulb. The merit of this high efficiency, however, would be somewhat diminished by the objectionable green hue. To give the impression of white we must also have the reds and blues, even though the eye is less sensitive to these shades.

So much for the quality of radiation. I now return to a brief discussion of the quantity. As we raise the temperature of a radiating surface we find that the amount of energy increases along with the march of its maximum toward the shorter wavelengths. Stefan first found, experimentally, the law connecting temperature with the amount of energy radiated. Later, Boltzmann gave it theoretical blessing. I venture to state the law here, because it is extremely simple. A "perfect" radiator (that is, one that is black when cool) will send out 37 watts for each square inch of its surface, at a temperature of 1000°. Now, if we double the temperature, the emission increases in the ratio of 2^4, that is, $2 \times 2 \times 2 \times 2$ or 16-fold. Similarly, a body at 6000° emits $6^4 = 1296$ times as much, or 48,000 watts per square inch. I have already given the corresponding value for the sun as 40,000. Hence the solar temperature, calculated from the quantity of its radiation, is just under 6000°, or 5800°, to be more exact. The agreement between this value and that derived previously from the color (spectral distribution) of the solar energy is altogether satisfactory.

It was necessary to define the "perfect" radiator, in the previous paragraph, because ordinary bodies do not always behave in so simple a fashion. Even carbon, which is one of the blackest substances known, departs to some extent from theoretical perfection. A tungsten filament shows a very appreciable defect of this sort. In consequence, one must introduce corrections to the observations, but I have purposely ignored them, in order to simplify the presentation.

⊙

The Earth's Atmosphere

One extremely troublesome correction remains, however, which I must discuss further; this is the problem of allowing for the absorption of the earth's atmosphere, through which we must make our solar observations. Many an astronomer has longed for an observing station on the airless moon, but none with more cause than the student of solar energy. And, for this reason, study of solar radiation outside the atmosphere is one objective of the current program of artificial satellites.

In entering the atmosphere, sunlight immediately begins to suffer depletion. At a height of 25 or 30 miles, it encounters a layer of ozone, which is practically impervious to the far ultraviolet. Virtually no radiation short of about 3000 angstroms gets through. We might just as well have a thick concrete ceiling above us, as far as this portion of sunlight is concerned. The molecular ozone, which consists of three atoms of oxygen tied together, is only one of the absorbent substances. Ordinary two-atom oxygen molecules take a heavy toll in the red and infrared. In the lower levels of the atmosphere, water vapor and carbon dioxide, along with many less important compounds, exact further levies. Add to these substances the dust, smoke, and the general scattering of light by all the constituents, and we face a very complex problem. Pure sunlight is badly distorted by the time it reaches our instruments.

The absorbed radiation goes to warm the upper atmosphere. The scattered portion flies in all directions, some out into space and some downward; it provides sky light. Since molecules scatter blue

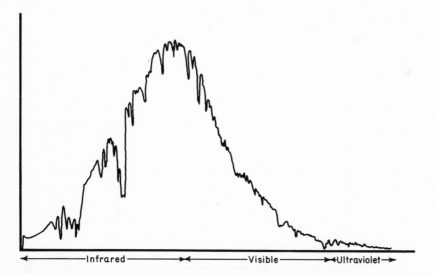

Fig. 42. Solar radiation, as measured by Abbot. The big dips, especially in the infrared, arise from absorption by molecules in the earth's atmosphere.

light more efficiently than red, the resultant color of this scattered radiation is blue. Hence the blue of the sky.

To escape from as many of these difficulties as possible, Dr. C. G. Abbot, chief authority on determination of the quantity of solar radiation, selected high-altitude desert sites for the observatories of the Smithsonian Institution. The desert gives the obvious advantages of clear skies, many observing days, and freedom from water vapor, smoke, and possibly dust. Also the general flatness and monotony of the surroundings ensure that the air at any given level above the ground will be practically uniform, a very important condition.

The fundamental observations are similar to those previously described. A spectrograph with a prism of rock salt (glass absorbs the heat rays and hence cannot be used) forms a spectrum. Abbot uses the equally acceptable *vacuum bolometer* instead of a thermo-couple to record the intensity of sunlight at each wavelength. The bolometer relies on the changes in electrical resistance of a fine blackened wire as its temperature rises because of radiation falling upon it.

Forgetting for a moment the curvature of the earth's surface, consider observations taken at two positions of the sun: first, when it is in the zenith, and second, when it is 60° from the zenith (Fig. 43). The path length *OB* is exactly twice the path *OA* and all of the absorption effects are doubled. This doubling does not mean that, if the beam is reduced to two-thirds of its initial value along *OA*, it will be reduced to one-third along *OB*. The final intensity along the

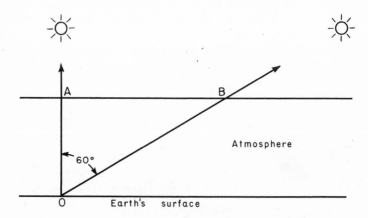

Fig. 43.   Absorption effects of the earth's atmosphere.

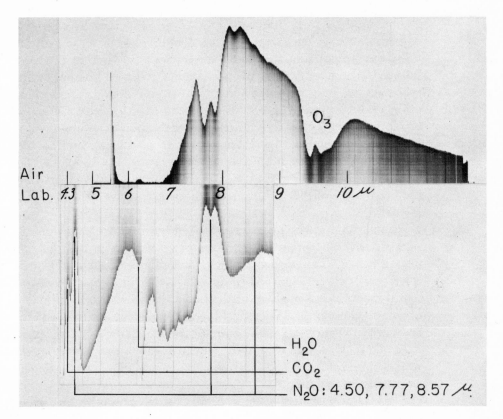

Fig. 44. Atmospheric absorption bands in the infrared: (*top*) solar spectrum; (*bottom*) labora-tory comparison spectra; the vertical axis indicates intensity of solar radiation. Absorption bands are marked for water (H_2O), nitrous oxide (N_2O), carbon dioxide CO_2), and ozone (O_3). (Arthur Adel.)

latter path will be ⅔ × ⅔ or ⁴⁄₉ of its value outside the atmosphere. If the intensity reduction along *OA* were one-half, that transmitted along *OB* would be ½ × ½, or ¼, and so on.

Now let us see what we can deduce from this fact. If we divide the measured intensity along *OB* by that along *OA*, the quotient is the transmission power for a vertical sun. Hence we may correct our measures and deduce the intensity outside the atmosphere. In practice, of course, we employ a more elaborate procedure, but the argument is similar. By studying each wavelength in turn we build up a composite curve of the solar spectrum as it would appear to an observer outside the earth's atmosphere.

The method fails, naturally, for those wavelengths that fail
entirely to reach the earth's surface. We have to fill these gaps by
using our imagination and best guesses. Sometimes, as in the near
infrared, where some molecule has taken a generous bite out of the
spectrum, we may guess the correction without introducing serious
error. Photographs taken from rockets, shot to altitudes of 100 miles
or more, indicate that the radiation in the region from about 2000
to 3000 A is somewhat less than we had originally expected on the
assumption that the sun's effective temperature was 5800° K. The
total value of the solar energy probably remains little affected by
the inaccuracy of this assumption, however, because the correction
is small.

An instrument, specially designed and built by McMath, multiplies by at least tenfold the astronomer's observing powers for the
near infrared. This device employs a lead sulfide photocell and
electronic recorder. For recording still farther in the infrared, the
sensitive receiver is made of lead telluride. The infrared studies by
McMath, Goldberg, Mohler, and Pierce have greatly improved our
knowledge of conditions in the atmosphere of both the sun and earth.

Large rockets, carrying spectrographs above the ozone layer,
which blanks out the solar ultraviolet, are adding new knowledge
about this interesting and important spectral region. Tousey and
co-workers of the Naval Research Laboratory, and others working

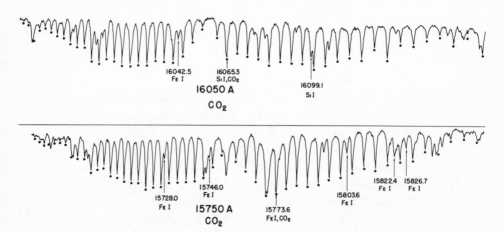

Fig. 45.   Solar spectrum, infrared, showing the CO_2 bands at 16050 A and 15750 A. The dots
indicate lines in the CO_2 systems. A few atomic lines are indicated. (McMath-Hulbert
Observatory.)

through the Air Force Cambridge Research Center, have secured a number of spectra that extend to about 2200 A. Many strong atomic lines, the most important of which belong to Mg II, Mg I, Si I, and Fe II, appear in this region. Analysis is difficult because the region is so rich in spectral lines and because the size and weight limitations as well as the shortness of available exposure for a rocket-borne instrument restrict us to the use of highly compressed spectra. Recent experiments have recorded the spectrum to 1000 A, including two strong hydrogen radiations, first discovered by Lyman, which lie at 1215A and 1025A, and several lines of carbon and silicon in various stages of ionization, all in emission. Rockets and satellites also indicate the presence of x-ray energy greatly in excess of that to be expected from a black body at 6000°K.

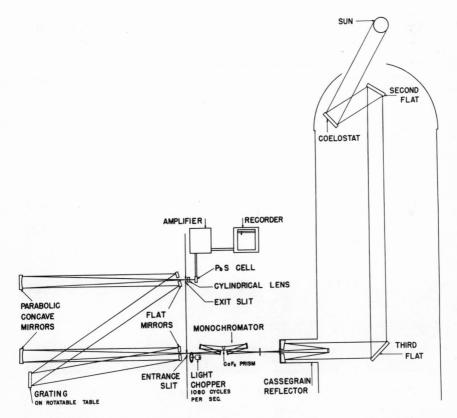

Fig. 46. Schematic drawing of optical system of a solar telescope and infrared spectrometer. The sensitive recording element is a lead sulfide (PbS) cell. (McMath-Hulbert Observatory.)

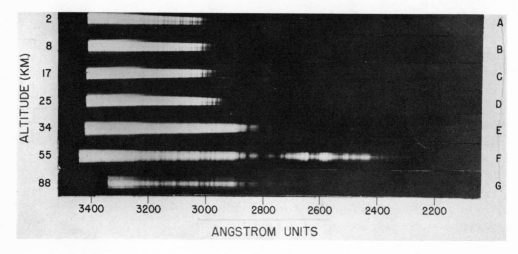

Fig. 47.   Solar spectrum photographed from a V-2 rocket on 10 October 1946. At the lowest levels, the ozone cuts the spectrum off sharply around 3000 A. At 34 kilometers (20 miles) the rocket begins to penetrate the ozone layer. At 55 kilometers (35 miles) the rocket is above the ozone and we glimpse for the first time the spectrum of the far ultraviolet, of which the most pronounced feature is the strong absorption of magnesium at 2800 A. Rockets rising to altitudes greater than 100 miles reveal additional features of the extreme ultraviolet region. (R. Tousey, U. S. Naval Research Laboratory.)

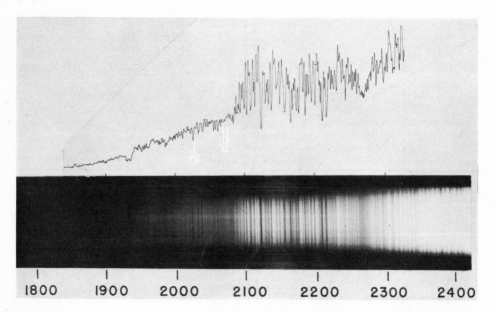

Fig. 48.   "Rocket ultraviolet" spectrum, 2400 to 1800 A. (R. Tousey, U. S. Naval Research Laboratory.)

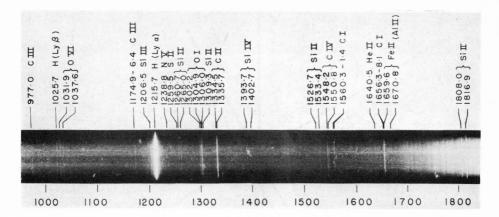

Fig. 49.   Solar spectrum recorded at an altitude of 115 kilometers above New Mexico on 21 February 1955. (U. S. Naval Research Laboratory.)

Fig. 50.   Composite of several exposures, showing complete solar ultraviolet spectrum from 3000 to 1000 A. (R. Tousey, U. S. Naval Research Laboratory.)

In the foregoing, I have used the *watt* to define the rate of radiation of energy. I might have employed other terms instead, for example the *horsepower,* which is 746 watts. The *calorie* is the energy necessary to raise the temperature of 1 gram of water 1 degree centigrade. One calorie per second is equal to 4.18 watts. Energy is the same, whether it be electrical, mechanical, or thermal. And we may readily transform energy from one form to another.

As an alternative method for finding out the amount of solar energy, we may measure the heating effect of the sun's rays directly. The instrument, which we term a *pyrheliometer,* comes in a variety of forms. The simplest is a hollow copper "pill box," blackened on the surface to make it absorptive to the radiation, and filled with water. When this device is exposed to the sun, the absorbed radiation causes the temperature of the water to rise, as recorded on a simple thermometer. Thus we obtain a measure of the energy in the sunlight. Such an instrument was first used by Pouillet, in 1837.

The so-called silver-disk pyrheliometer contains a blackened silver

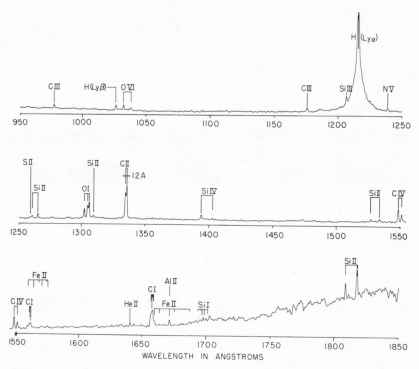

Fig. 51.   Trace of the intensity of the solar radiation at various wavelengths in the rocket ultraviolet, 21 February 1955; see Fig. 49. (U. S. Naval Research Laboratory.)

disk to receive the energy. Silver, the best available conductor of heat, minimizes the delay in the warming of the bulb of an L-shaped thermometer inserted in one edge of the disk. Such an instrument, as designed by Abbot of the Smithsonian Institution, is shown in Fig. 52. The rays pass down the hollow protecting tube to the silver disk at the bottom. The long stem of the L of the thermometer, which is readily accessible to the observer, is contained in the tube at the top. The entire apparatus is mounted with a clockwork drive that keeps it pointed sunward.

The most accurate modern pyrheliometers are ordinarily placed in a sort of thermos bottle, which reduces the heat losses. In some forms, water circulates continuously instead of remaining static, and delicate electrical devices record the rise in temperature. All of the pyrheliometer readings require correction for the transmission of the earth's atmosphere.

I must briefly mention one other instrument, the pyranometer, which appears in Fig. 53, ready for action. Within the transparent hemispherical cap lies a specially designed thermocouple. The circular disk, projecting far beyond the receiving surface, casts a shadow on the couple, so that only the sky radiation beyond the edge of the solar disk records.

What advantage do we gain from use of this instrument? The bolometer and pyrheliometer introduce two difficulties which I have previously only mentioned. The theory of evaluating the atmospheric correction from measures of the sun at different altitudes

Fig. 52.   Silver-disk pyrheliometer. (Abbot.)

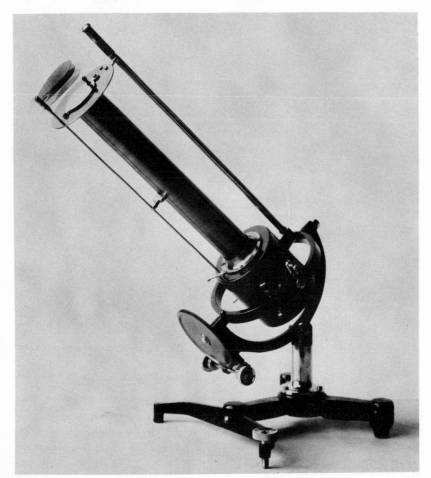

implies that the air remain uniform throughout the entire day. For the same reason, we cannot use days that are only partly clear.

Abbot has found, however, from simultaneous records by all methods, a close correspondence between sky brightness and atmospheric transmission. You may even have noted the effect yourself. The sky is much brighter when haze is present than when it is

Fig. 53.   The pyranometer. (Abbot.)

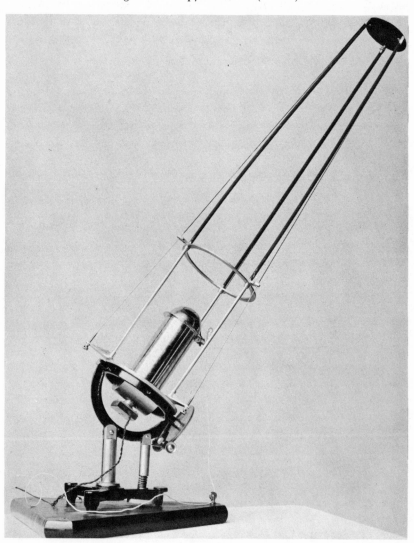

absent. Consequently, a single pyranometer reading provides an estimate of sky transparency and affords a determination of the amount of solar radiation. Further, the labor involved in the taking and interpretation of the observations is enormously lessened.

From years of observational study, Abbot gives for the *solar constant,* as we ordinarily term the quantity of solar radiation incident outside the earth's atmosphere, the value of 1.94 calories per square centimeter per minute, or about 1.5 horsepower per square yard. Recently F. S. Johnson has calculated improved corrections for the ultraviolet and infrared atmospheric absorptions. With these, Johnson obtains a value of 2.00 calories for the solar constant.

The existence of variations in the amount of the solar radiation is a question of much importance, and of long-standing controversy. Variability of the sun's radiation, if real, is highly important, because of its possible effects on the earth. In view of the very marked changes of the solar surface over a sunspot period, we should not be surprised to find a variation. Indeed, we may well be astonished that the sun varies so little. Various statistical studies have suggested a periodicity in the measures, a gradual rise and fall with an 11-year cycle, apparently coincident with the sunspot variations; we receive more energy when spots are numerous.

However, the most recent statistical studies, by Sterne and Mrs. Dieter, indicate that the solar constant shows no systematic or periodic variations larger than 0.2 percent. Further, they find no periodicities common to the simultaneous observations from the two Smithsonian stations, at Table Mountain, California, and Montezuma, Chile.

Perhaps the best method for determining the reality of the supposed fluctuations in the solar constant is by accurate measures of the brightness of the planets. Since these bodies shine by reflected sunlight, they should mirror any changes in the brightness of sunlight. One might expect variations arising merely from spots or inequalities of its surface and the planet's rotation. During a short period of observation, Stebbins at Lick Observatory found the magnitude of Uranus to be surprisingly constant. Hardie and Giclas, at the Lowell Observatory, found from study of Uranus and Neptune that any variations in the output of solar energy, in the five years 1949–1954, are probably less than 0.4 percent. H. L. Johnson and B. Iriarte, continuing this work, have announced an increase of about 2 percent in the solar constant from 1954 through 1958.

While the latest studies indicate the solar radiation to be remarkably constant—indeed, no other star has been proved to be as constant as the sun!—we must remember that the data apply only to a limited portion of the solar radiation. From the variable electrification of our own upper atmosphere, we infer that significant changes do occur in the far ultraviolet. But here we cannot make direct observations because the ozone is so completely opaque. It is expected that the artificial satellite program will greatly extend our knowledge of this region of the solar spectrum, by observations outside the earth's atmosphere.

For the present we shall tentatively admit the likelihood of fluctuations in the solar ultraviolet radiation. In the final chapter of the book I shall have occasion to discuss possible terrestrial effects of these variations.

5

Solar Chemistry

In a previous chapter, I have told you something of the mad dash of atoms as they struggle for the symmetry of perfect chemical union. This ability of atoms to cleave to one another and form stable molecules is the reason for the existence of chemical science. I might even say that the struggle itself is chemistry.

Here on the earth the atoms are moving at leisurely speeds. For example, a supermicroscope would show the molecules of our air traveling at average velocities of some 14 miles per minute. It may seem paradoxical to refer to such a speed as "leisurely." I use the term in the relative sense. In the sun's atmosphere, atoms move with velocities of from five to ten times greater than on earth. The higher speeds, of course, are due to the higher temperature on the sun. As a matter of fact, the velocities themselves, or, more exactly, the energies of the moving atoms, determine the temperature. When we heat water on the stove, we are merely imparting to the molecules greater speeds, greater energies. If we could speed them up by other means, the temperature would rise similarly.

If we could see the molecules of our atmosphere in motion, we should be astonished at the frequency of collisions between the particles. Moving at random, with no driver to guide them, the molecules crash recklessly into one another. Fortunately, the elastic cloud of electrons that surrounds each atomic nucleus acts as a very effective "bumper." Rarely does collision damage an air molecule. The particle bounces away and dashes off on some new path until another impact deflects it again. Each molecule of air, in the course of a single second, collides about 10,000 million times!

At higher temperatures, however, the atoms and molecules, moving more rapidly, are not always so fortunate. Two automobiles, traveling at five miles an hour, may collide without damage. But increase the speed to fifty—and fenders fly in all directions. The energies increase with the *square* of the velocity. A collision at fifty miles an hour is not ten times but a hundred times more serious than one at five miles an hour—a simple fact that the driving public seems not fully to appreciate.

In the atmosphere of the sun, the atoms will sometimes be dented, at other times broken, by the impacts. A hydrogen atom, hard hit by another atomic projectile, may change from its most common vibration pattern to some other type (Fig. 36). Atoms, however, unlike automobiles, possess the ability to repair themselves. A fraction of a second later the atom sends out a quantum of radiation, and reverts to its initial stable form, ironing out the "dents."

Occasionally the crash is so terrific that the atom may lose its electron entirely. When this sort of catastrophe happens, the repair is not immediate. The atom, now an ion, must wander on until it encounters another free electron. And then the repair occurs automatically. Nor does a hydrogen atom have to worry about the source of its new electron. Electrons are all alike; as spare parts, they fit any atom.

Collisions are not the only means for denting and disrupting atoms. I have already indicated that radiation plays a similar role. An atom, absorbing and emitting light, changes from one vibrational pattern to another and back again. If the disturbing quantum has enough energy, that is, if it has a sufficiently small wavelength, it too can ionize the atom. In the sun both processes are active. We see in the spectrum of the sun lines of both neutral and ionized atoms.

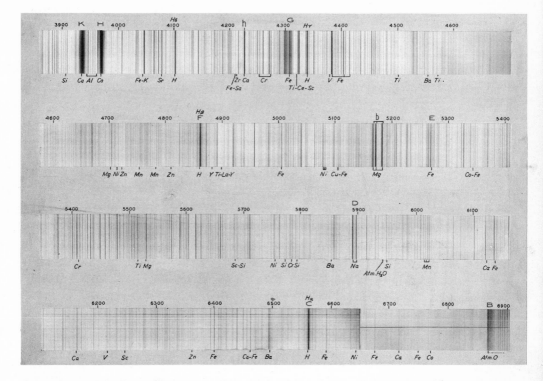

Fig. 54. Mount Wilson Observatory photograph of the visible solar spectrum, from violet (3900A) to red (6900A). (Courtesy *Sky and Telescope*.)

The spectrum of the sun is essentially *continuous,* that is, all colors of light are present, from the far ultraviolet through the visible rainbow band of color and on into the infrared. A careful examination, however, shows that certain colors are missing (Fig. 54). As mentioned in Chapter 3, the entire spectrum contains thousands of dark lines of different intensities, termed the *Fraunhofer lines* after their discoverer. Some, called the *telluric lines,* arise from absorption by various molecules in the earth's atmosphere. The sun's gaseous envelope produces the remainder. We shall concern ourselves in this chapter only with the latter.

The mode of origin of these dark lines was long a puzzle to astronomers. They clearly corresponded in position to the bright lines of the various chemical elements. Eventually, astronomers recognized that the missing radiations were *absorption lines;* the atoms in the cooler, outer layers of the star absorbed, from the continuous radiation emanating from the lower depths, the specific colors appropriate to the chemical nature of the various atoms.

We still do not understand completely how the continuous spectrum arises. [This question is discussed in L. Goldberg and L. H. Aller, *Atoms, Stars, and Nebulae* (Harvard University Press, Cambridge, 1950), Chap. 6.] Deep in the stellar interior, where the density is high, the crowded atoms jostle one another so violently that they rarely have the chance to assume vibrational patterns characteristic of absorption-line formation. Most of the absorption and emission is produced by the flying electrons as they switch from one path to another, occasionally attaching themselves momentarily to some ion, or even to some neutral hydrogen atom, thus forming a negative hydrogen ion.

The "crowded" conditions in the sun's atmosphere arise not so much from the numbers of atoms as from the high speeds at which they are moving. A church in which 500 people are sitting more or less quietly may not seem crowded at all, whereas 50 couples jitter-bugging on a dance floor of equal size may feel very cramped for room as they frequently collide with one another.

From the standpoint of atomic population, the density of the sun's atmosphere is far less than that of the earth's. Since the sun is gaseous throughout, the word "atmosphere" needs further clarification. We usually refer to the apparent boundary of the sun, which produces the continuous spectrum, as the *photosphere,* the sphere of light. In the strictest sense, the photosphere is not a sharp boundary but a region. A single thickness of fine gauze may impair but slightly the vision of objects behind it. But a thousand pieces of gauze may be as opaque as a wall. Looking into the atmosphere of the sun is analogous to looking into a pile of gauze. Electrons interacting with hydrogen atoms to form negative hydrogen ions furnish a major source of opacity in the infrared, while recent work by Varsavsky indicates that the H_2 molecule may be important in the ultraviolet. We see down into gaseous layers where the material becomes substantially opaque to our vision. The portion lying above the zone of complete opacity comprises what we usually call the solar atmosphere.

The pressure in the photospheric layers, as I have said, is not high; probably it is of the order of 10 percent of the atmospheric pressure at the surface of the earth. Making due allowance for the greater gravitational force of the sun, we estimate that the mass of atmosphere above each square yard of solar photosphere is only about 50 pounds. The earth, by way of comparison, has about 15 pounds of air over each square inch, or 10 tons per square yard. The total mass of the sun's atmosphere is about 100 thousand million million (10^{17}) tons. Large as this figure seems, it is, nevertheless, only about 20 times larger than the total mass of the air that surrounds the tiny earth.

A very simple experiment confirms our conclusion that the sun's atmosphere is indeed very thin. Dissolve a pinch of salt (sodium chloride) in a glass of water. Heat a wire in a gas flame until all trace of color disappears from the flame. Dip the wire into the salt solution and again introduce it into the flame. A brilliant yellow flare, characteristic of the element sodium, results. From the nature of our experiment we know that very little sodium can be in the flame. And yet if we pass white light, say from an incandescent lamp, through the flame, we observe, with a spectroscope, absorption lines far more intense than those of the solar spectrum. The experiment shows that the amount of sodium in the sun's atmosphere is very minute; it further exemplifies the great sensitivity of the spectroscopic method and demonstrates the high absorptivity of atoms for light of the same wavelengths that they can emit.

From laboratory studies of the spectra of various elements and from theoretical calculations of the ways in which atoms can absorb and radiate light, we attempt to interpret the observed solar spectrum. We find what elements are present and in what amounts. Since the computations and theory are fairly technical matters, I shall not detail them here. The intensity of a spectral line, however, is not of itself a direct gauge of the quantity of material present. The atom has a natural tendency to absorb certain colors more readily than others. Also the temperature has a marked effect on the spectrum. There are, for example, thousands of iron lines in the spectrum of the sun. They range in intensity from among the strongest to the weakest in the spectrum. If the temperature of the gas is low, most of the atoms are in their simplest vibrational pattern. The number of colors that such an atom can emit or absorb is limited and the resultant spectrum will be simple, consisting of but a few lines. The higher the temperature, however, the greater the excitation and the more atoms we shall find in other vibrational patterns; the spectrum will be correspondingly more complex.

The astronomer puts this fact to practical use. He compares the intensities of the lines that appear only under conditions of high excitation with those of the ordinary lines. The differences in intensities enable him to gauge accurately the temperature of the source, which is, in this case, the sun's atmosphere. The temperature determined independently by several investigators from studies of the dark-line Fraunhofer spectrum is about 4800°K (8100°F).

This value is about 1000°K lower than that deduced from the quantity of sunlight, as explained in the previous chapter. The figure of 5800°, however, refers to the effective radiating solar surface, i.e., the photosphere, whereas the lower value of 4800° is appropriate to the surrounding atmosphere wherein the dark lines are produced, a region known technically as the *reversing layer*. We should not be surprised to find the atmosphere somewhat cooler than the base, although the 1000° differential was greater than most astronomers expected. But more of this later.

Chemical Composition of the Solar Atmosphere

⊙

One of the goals of solar studies is the determination of the chemical composition of the sun's outer layers. The most intense lines in the visible solar spectrum arise from atoms of ionized calcium, the so-called H and K lines. This observation does not mean that calcium atoms, minus one electron, form the major part of the sun's atmosphere. We have already seen that line intensity alone is not a perfect index. We must correct for temperature, for the intrinsic atomic absorption power, and so on.

It then turns out that hydrogen, whose lines are much weaker than the calcium ones just mentioned, is actually many thousand times more abundant. The great majority of calcium atoms in the atmosphere are in just the state from which they can absorb the violet H and K lines. However, only one hydrogen atom in a million —we do not know the exact figure—is ready to absorb visible radia-

Fig. 55.   Spectrum of A0 star with well-developed hydrogen series.

tion. To do so, a hydrogen atom must have previously been excited to a high energy state. And at 5800°, not to mention 4800°, the radiation and force of collisional impact are too feeble to so excite many hydrogen atoms. Only in the hotter A0 stars, like Sirius or Vega, do the hydrogen lines reach a truly enormous intensity (Fig. 55). Helium lines show in stars with still higher temperatures, and ionized helium in the hottest of all, for helium is one of the toughest atoms to dent or to break apart.

Although temperature plays the major role in breaking the atom to pieces, tearing away the outer electrons, in the repair or inverse process pressure also has an important part. An ionized atom must remain disrupted until it chances to meet and capture some wayfaring electron. And, clearly, its chances of finding the spare part are much greater in matter of high than in matter of low density. For a given temperature, atoms show a higher percentage of ionization under low than high pressure. Conversely, if we know the temperature and if we can further estimate, from the intensities of spectral lines, the numbers of neutral and ionized atoms, we can calculate the pressures.

The value previously quoted, one-tenth of an atmosphere, refers to the neutral hydrogen gas near the sun's surface. The atmospheric density gradually decreases with height above the solar surface as the photosphere fades into the overlying chromosphere. We see the latter regions most clearly at the time of total solar eclipse, when the moon has completely covered the photosphere. For a brief moment the chromosphere flashes into view, its spectrum consisting of bright lines, appropriate to the tenuous gas. A spectrograph records thousands of crescent-shaped patterns of emissions, each corresponding to a different wavelength (Fig. 56).

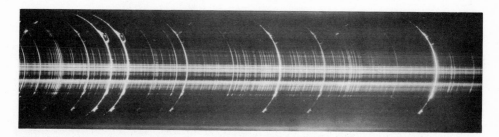

Fig. 56.   Flash spectrum of solar chromosphere and prominences, taken during the eclipse of 31 August 1932 at Fryeburg, Maine. (Lick Observatory.)

This *flash spectrum,* as the phenomenon is usually termed because of its brief duration, bears a marked resemblance to the ordinary Fraunhofer absorption spectrum. There are, however, numerous differences, chiefly in the intensities of the lines of the ionized metals, which appear relatively stronger in the chromospheric layers. The lower pressure, as previously explained, is conducive to increased ionization.

Low density, however, is not alone the answer to the peculiarities of the flash spectrum. A preview of certain difficulties associated with any simple picture of the solar atmosphere is appropriate here. In the chromospheric spectrum the lines of the hard-to-excite atom, neutral helium, are very intense, although they show up rarely, and then only weakly, in the Fraunhofer dark-line spectrum.

A relatively simple calculation shows that the observed degree of helium excitation can arise only by the action of either intense ultraviolet radiation or fast-moving electrons. The ordinary 5800° solar temperature could not possibly produce the required excitation conditions. A temperature of 20,000° or 25,000° is indicated for the helium lines, whereas one of 4000° is adequate for various lines of the metals. In an attempt to account for the two extremes, Athay and Menzel have suggested that the chromosphere consists of alternately hot and cool columns of gas. But this single problem is just one of many similar difficulties yet to be encountered in our attempts to analyze the complete data and to build therefrom a satisfactory picture of the structure and activity of the sun.

To interpret the intensities of the absorption lines in the solar spectrum, we pose ourselves the following problem. Assume that we have only the photosphere and no atmosphere at all. The spectrum would then be purely continuous—unbroken by dark lines. Now let us build up an atmosphere atom by atom. How shall we expect the absorption lines to "grow"?

A given line of the spectrum is not strictly monochromatic, that is, single colored. Owing partly to an intrinsic breadth and partly to their motions, atoms producing the line will absorb effectively over a small range of color. An atom moving toward the observer will absorb light in the short-wavelength or "violet" edge or "wing" of the line; one moving away from the observer will absorb in the long-wavelength or "red" wing. The Doppler effect, previously discussed, is the cause of this phenomenon. Thus high temperature widens spectral lines, because of the large velocities of the moving atoms.

Iron atoms in the reversing layer, absorbing like a perfect oscillator in a miniature radio set, will produce a barely perceptible absorption line with about 1 billion (10^9) atoms per square centimeter. Only about 1 percent of the light will be cut off over a narrow region. Ten billion (10^{10}) atoms will absorb something like 10 percent over the same region, whereas 100 billion (10^{11}) atoms will produce a very black line indeed—almost completely absorbed at the center.

If, now, we increase the number of atoms by an additional factor of 10, the absorption line can grow only a little stronger. The line is *saturated;* in other words, it is already so black at the center that most of the further absorption consists of a slight widening.

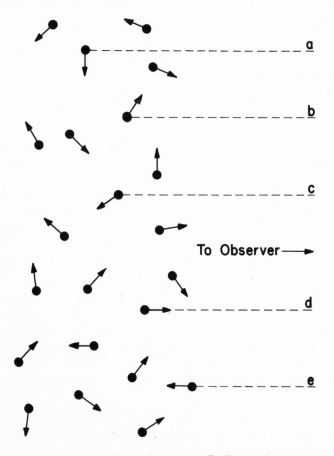

Fig. 57.   Random motions of radiating atoms. Radiation from atoms *c* and *e* shifts toward the red, while that from *b* and *d* moves violetward. The wavelength of the radiation from *a* is unaltered.

We measure the absorption in terms of the *equivalent width*, the amount of energy that would be subtracted from the spectrum by a completely black, rectangular line of width W angstrom units (see Goldberg and Aller, chap. 6). We term a graph of W against the number of atoms the *curve of growth*. The theoretical curve (Fig. 58) increases sharply for small values of N, the number of atoms. Then it flattens off for a piece, finally turning upward again.

The details of this curve are of primary interest only to the professional. The theoretical development is due to many investigators, but the original work was started by the Dutch astronomers, Minnaert and Slob. This curve has proved to be basic in the interpretation of solar and stellar spectra. It is fundamental for quantitative analysis. One measures the W. For example, we may find some line in the solar spectrum whose W is 1 angstrom unit. Reference to the curve will show that some 10^{14} atoms per square centimeter are required to produce a line of this intensity at wavelength 6000 A.

Of course the problem is not quite as simple as depicted above.

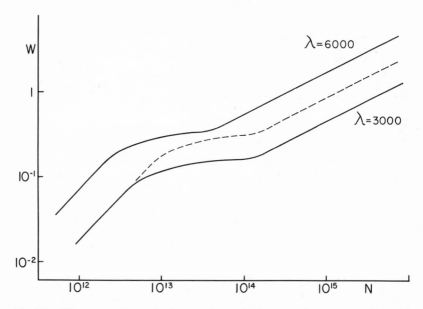

Fig. 58. The curve of growth. A semischematic diagram showing the relation between equivalent width W of a spectral line and number of absorbing atoms N, for two different wavelengths. The dashed portion of the line corresponds to a higher thermal temperature of the gas. By plotting log W/λ (λ = wavelength) instead of log W, we can make the two separate curves coincide.

Fig. 59.   M. Minnaert, Dutch astrophysicist.

Atoms rarely act like the assumed perfect oscillators. We must first correct the value for the true atomic absorbing power in a given line. And then we must introduce further corrections for temperature. The final result is a spectroscopic determination of the chemical composition of the solar atmosphere.

The first comprehensive determinations of solar abundances were made in 1929 by Russell of Princeton, who employed Rowland's estimates of intensities, in lieu of the curve of growth. The latter is more accurate but was not available when Russell did this pioneer work.

It is of interest to compare the results with the chemical composition of the earth's crust, as shown in Fig. 61. The proportions of the heavy metallic elements correspond fairly well in the two bodies. But the lighter elements seem to be very much more abundant on the sun. The most outstanding discrepancies are for hydrogen and helium. True, the solar abundance of the latter is quite uncertain. The helium value is particularly sensitive to small errors in the temperature. If we disregard helium completely, we may say that

Fig. 60. Henry Norris Russell. (Photograph by Orren Jack Turner.)

the sun's atmosphere consists of chemically pure hydrogen. We repeat: the solar gases are chemically pure, but not spectroscopically pure, hydrogen.

There is a great distinction between these two degrees of purity, as the chemist well knows. So sensitive is the spectroscope to minute contamination that a chemically pure sample may cost, say, 10 cents a pound, whereas a spectroscopically pure sample of the same substance may run to $500 an ounce.

In this surprising chemical difference between earth and sun we see the start of many questions. The fact that hydrogen appears to predominate throughout the universe, as well as in the sun, further stresses the anomaly. Something must be wrong with our earth.

The explanation may hold a clue to evolutionary trends. In the primordial scheme, when the earth was forming, perhaps it had a normal amount of hydrogen. Then, still at an early stage, the lighter elements disappeared, somehow evaporated into space [see F. L. Whipple, *Earth, Moon, and Planets* (Harvard University Press, Cambridge, 1952), chap. 14]. Our moon has lost its atmosphere completely. So has the planet Mercury. Mars has only a vestige of

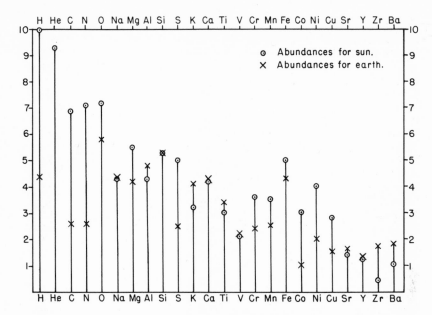

Fig. 61.   Comparative abundances of elements in the earth and the sun. The circles indicate abundances for the sun, and crosses abundances for the earth. The abundances refer to numbers of atoms and not to total weights of the material. The absolute scale of each set of abundances is arbitrary. The scale on the left-hand column is logarithmic; in other words, the number 7 means 10^7 atoms, whereas the number 4 means only 10^4 atoms. Solar abundances are from the work of Minnaert, Hunaerts, and Unsöld. The terrestrial values are from the work of Washington and Clarke and refer only to the earth's crust.

surrounding gases. The process has been a sort of distillation, in which the less massive planets have lost at least the lighter gases.

In apparent harmony with this view stands the fact that Jupiter, Saturn, Uranus, and Neptune appear to have retained most of their original hydrogen. The spectroscope reveals ammonia (NH_3) and marsh gas (CH_4) as abundant constituents of the atmospheres of the giant planets.

To confound our theories, however, Kuiper has shown that Titan, the great satellite of Saturn, also has an ammonia-methane envelope. And Titan is no larger than Mercury, which has lost all of its atmosphere! Of course Mercury is close to the sun, whereas Titan is far away, in the cold depths of space. But if ever the satellite was molten, it should have lost the hydrogen it now appears to possess.

An alternative hypothesis, suggested by Kronig, is that the earth's hydrogen has disappeared, not into space, but into the interior. Hot

TABLE 1.
ABUNDANCES OF ELEMENTS IN THE SOLAR ATMOSPHERE.

	Percentage volume	Mass (mg/cm²)
Hydrogen	81.760	1200
Helium	18.170	1000
Carbon	0.003000	0.5
Nitrogen	.010000	2.
Oxygen	.030000	10.
Sodium	.000300	0.1
Magnesium	.020000	10.
Aluminum	.000200	0.1
Silicon	.006000	3.
Sulfur	.003000	1.
Potassium	.000010	0.003
Calcium	.000300	.2
Titanium	.000003	.003
Vanadium	.000001	.001
Chromium	.000006	.005
Manganese	.000010	.01
Iron	.000800	.6
Cobalt	.000004	.004
Nickel	.000200	.2
Copper	.000002	.002
Zinc	.000030	.03

From Goldberg and Aller, *Atoms, Stars, and Nebulae* (Harvard University Press, Cambridge, 1950).

rock, under high pressure, can absorb vast amounts of hydrogen with no perceptible increase of density.

The astronomer, with the entire universe open to scientific exploration, can only sympathize with the terrestrial physicist and chemist, whose horizons are so limited. This minute speck of dust we call the earth has one mark of distinction—its anomalous situation with respect to the chemical and physical state of the rest of the universe. Not only is its composition unrepresentative. Its temperature conditions are likewise exceptional. They are intermediate between cold and hot.

Most of the matter of the universe is in stars and possesses a high temperature, caused by release of nuclear energy in the stellar interiors. An appreciable fraction, thinly strewn in the great spaces between the stars and galaxies, is extremely cold, only a few degrees above the absolute zero.

To the peculiar position of the earth in the cosmic temperature scheme we owe the greatest anomaly of all—the presence of life on its surface. I mention the subject here because life is also a form of solar chemistry. The mysterious reagent, chlorophyll, snatches energy from a beam of sunlight and locks it up in the growing plant cells. Photochemistry—the chemistry of light—makes life possible.

But science as yet has given no answer to the deepest question of all: where and how did life get its start? Fossil records in the rocks depict the broad evolutionary sequence from protozoa to man. The detailed processes of development are obscure, but certainly the forces of solar radiation played a part.

If life started like a spontaneous combustion, forces of chemistry must have provided the impetus required. We do find some of the necessary ingredients—water, ammonia—occurring naturally. Perhaps, in the warm seas of a prehistoric era, over a period of time so extended that even the improbable had a chance of occurring, there grew an enormous molecule of protoplasm—enormous, that is, in terms of minute atomic diameters. Urey, of the University of Chicago, has recently proposed a plausible scheme in terms of specific chemical processes.

And then the most mysterious event of all occurred. Energized by an ultraviolet quantum or perhaps by a cosmic ray, the molecule

Fig. 62.   A. Pannekoek, Dutch astrophysicist, well known for studies of physical conditions in the atmospheres of sun and stars.

suddenly began to draw other material into itself. Using itself as a mold, matching atom for atom, the animalcule formed a duplicate of itself and, eventually, split in two, thereby initiating the greatest and most complex of all chain reactions—biological reproduction.

Since I was not present at the event above described, I cannot guarantee it to be correct in every detail. The event occurred perhaps 1.5 billion years ago. For many millions of years the single cell had its way, until perhaps 550 million years ago, when the wormlike invertebrates crawled on the scene, to be followed some 150 million years later by the fishes. Frogs and toads came about 300 million years ago. Then the dinosaurs and other great reptiles dominated the picture from about 200 million to 60 million years ago, only to vanish with the evolutionary rise of the mammals.

The most primitive forms of man date back perhaps 1.5 million years, and modern man, *Homo sapiens,* 25,000 years or so. Thus, we —the human race—have occupied the stage only 1/50,000 of the entire time that the drama of life has been playing. If the whole performance were compressed to occur in a conventional 3 hours instead of 1.5 billion years, the proportionate time allocated to man would be only one-fifth of a second. In our own estimation, we are stealing the show. Is it not a trifle early to tell?

6

Sunspots—Magnetic Islands

Sunspots radiate less light and heat than their surroundings (see Chapter 2). The centers thus appear relatively dark. But do not imagine for a moment that the spot is completely black. A spot emits about 10 percent as much energy as an equal area of the solar surface. In visible light the percentage is still smaller. Even so, if I could, with the wave of a wand, cause the sun to vanish except for a single large spot group, the earth would by no means be plunged into darkness. The spot would shine with a brightness equal to 100 full moons. Its radiation alone would still dazzle the unprotected eye.

Individual spots differ markedly in size and duration. We frequently detect small "pores," a few hundred miles across, near the limit of telescopic resolution. If we could employ larger and more powerful telescopes we should probably detect even smaller spotlike disturbances. But the major interest for us lies in the larger spots and spot groups that form and dissolve on the solar disk. What is their nature? How do they come into being? Above all, what forces are responsible for the 11-year cycle of variation in the number of spots?

The number of spots varies from day to day and year to year. Eleven years, on the average, elapse between times of maximum spottedness, with a well-defined minimum in between, when many days may pass without a single observable blemish on the sun's surface.

The blackness of a spot indicates that it must be cooler than surrounding portions of the solar disk, but this is not the only argument for a lowered temperature. For example, we see lines in the spectra of spots that are due to molecules. At the higher temperatures of the solar reversing layer, these molecules are fully dissociated—broken up into their component atoms. Also, in the spots the low-temperature lines of neutral metals are strengthened; the lines requiring higher excitation are weakened. The various observations indicate that a spot possesses a temperature of about 4000°C, approximately 2000° less than that of the solar photosphere.

The dark center, or umbra, of a spot contrasts sharply with the filamentary, much brighter penumbra. The outer edge of the penumbra is fairly sharp, but irregular. Sometimes bright streaks of light bridge the spot, apparently splitting the umbra into two parts. The larger spots often display such activity. This phenomenon probably results from the presence of bright prominence material, which will be discussed in greater detail in Chapter 7. The umbra

Fig. 63.   Filamentary structure of spot penumbrae, the great spot group of 6 April 1947 (see Fig. 18). Note the ragged edges of the penumbral material and the luminous bridges tending to cross the two larger spots. Granulation of the solar surface is clearly shown. (Mount Wilson Observatory.)

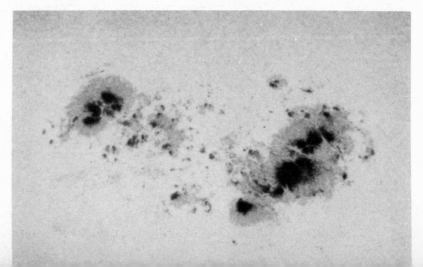

is not actually divided. The luminous bridge lies at a much higher atmospheric level than the spot.

⊙

Records of sunspot activity go far back into history. The Chinese have mentioned "flying birds" seen on the solar disk and the descriptions clearly indicate that these markings were large, naked-eye groups. To be so visible, the spot or spot group must be at least 25,000 miles or so in diameter. A number of such naked-eye groups appear during each sunspot maximum.

The earliest Chinese records of naked-eye spots were in the year 28 B.C. Sarton has listed a number of references, published between the years 807 and 1633. A dark spot that appeared about 807 was supposed to have presaged the death of Emperor Charlemagne seven years later.

Most of those who observed spots in this medieval period attributed their cause to transits of the planets Mercury and Venus across the solar disk. Even Kepler, who should have known better, explained spots observed 18 May 1607 as a transit of Mercury. Thus did ancient beliefs concerning the purity of the sun distort the thinking of early observers.

Galileo made the first truly scientific solar observations with the telescope and recognized that spots must be a phenomenon of the

Fig. 64. Sun at maximum and minimum of spot activity. On 30 November 1929, there were many spots; on 22 June 1931, there were no spots. (Mount Wilson Observatory.)

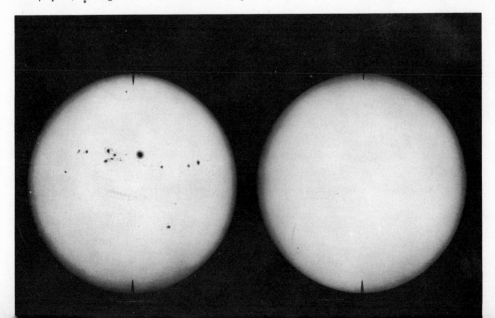

solar surface. He noted that they formed and dissolved on the face of the sun. Moreover, he discovered that they progressed regularly across the disk, a property he correctly interpreted as a result of solar rotation.

However, the sun does not turn on its axis like a solid body. Its rate of rotation slowly diminishes from equator to poles. If we could line up a series of spots along the meridian and start them off, like runners in a race, the equatorial spots would win (Fig. 65). They complete a circuit once in about 25 days, whereas those at 40° require about 27 days.

Spots are infrequently observed in latitudes above 35° and almost never beyond 40°. In August 1956, a period of high solar activity, a group of spots did appear at the unprecedented distance of 50° from the solar equator. The great majority of spots, however, occur between the latitudes of 5° and 30°, both north and south of the equator. The polar zones are unspotted. To determine the rotation period for higher latitudes, we rely on spectroscopic methods and the Doppler effect.

Table 2 gives the number of days required at various latitudes for one solar rotation. The values for spots are those of Nicholson and Miss Ware, of Mount Wilson Observatory, who employed only long-lived single spots for the purpose. Their values appear to be slightly

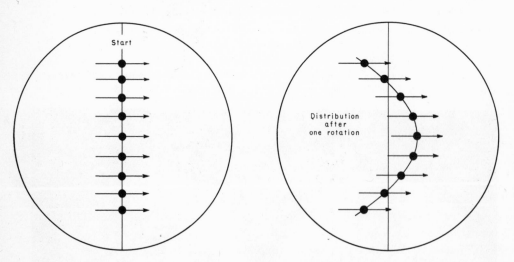

Fig. 65.   Schematic demonstration of solar rotation: (*left*) spots aligned along a meridian; (*right*) distribution after one rotation, with equatorial spots ahead of others.

TABLE 2.

TIME OF ONE SOLAR ROTATION AT VARIOUS LATITUDES.

Latitude (deg)	Spot days	Spectroscopic days
0	25.14	24.64
15	25.50	25.41
30	25.53	26.45
45		28.54
60		30.99
75		33.07

in excess of those measured by observers at Greenwich from long-lived spot groups. The spectroscopic measures were made by Adams at Mount Wilson.

⊙

The Sunspot Cycle

At the beginning of the 11.2-year cycle, most of the spots lie in latitudes of 20° to 30°. As the cycle progresses, the spot zones drift slowly toward the equator, a law discovered by R. C. Carrington about a hundred years ago, and traced by Spoerer back to 1621. Near sunspot minimum, one occasionally finds spots of the old and new cycles present simultaneously. The two cycles are readily distinguished not only by position on the disk, but by a characteristic magnetic effect, which will be discussed in detail later on.

The sunspot cycle is not exactly periodic. The average interval between maxima (or minima) is 11.2 years, but the actual values have ranged from 7.5 to 16 years. The ascent to maximum is usually more rapid than the decline, 5.2 compared with 6.0 years.

The variability in numbers of spots was noticed shortly after their discovery, but the cyclic character of the fluctuations was not recognized until 1843, when Schwabe made known the results of 17 years' study. Wolf, of Zurich, later took up the investigation and unearthed many old records of sunspots, which enabled him to extend the data back nearly to the original discovery in 1610. Wolfer, Brunner, and Waldmeier have subsequently carried on this basic work. Many observers, amateur and professional, now contribute to the study.

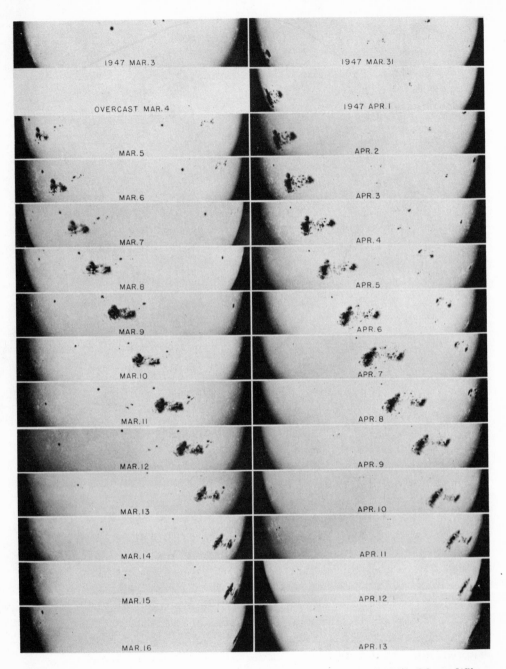

Fig. 66.   Solar rotation, shown by a large spot group, March and April 1947. (Mount Wilson Observatory.)

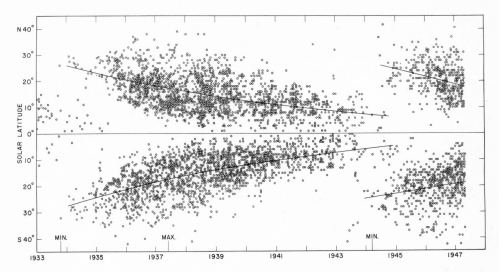

Fig. 67.   Spoerer's law of the distribution of sunspots in latitude. Note how the spot distribution moves equatorward during the development of a cycle. (*Sky and Telescope.*)

Wolf introduced the term *sunspot number* as a measure of solar activity. He defined the quantity as follows:

$$R = k(f + 10g),$$

where R is the sunspot number; g is the number of disturbed regions, either groups or spots; f is the total number of individual spots; k is a factor depending upon the observer and the size of his telescope.

Let us see how this definition works. You note, let us say, two groups each consisting of a pair of spots, and one isolated spot. Thus $g = 3$ and $f = 5$, which give $f + 10g = 35$. In order to relate your results to those of other observers, the individual values for a large number of days are intercompared with the data published from Zurich, where all records are sent for analysis. Your value of k is assigned by Zurich to bring your measures into best agreement with their standard.

This procedure is somewhat arbitrary. A large telescope may show more faint spots at one time than will a smaller one. Nevertheless, the means of many observers do have significance. Monthly averages are better than daily values and yearly means are still better. Table 3 lists the mean annual spot numbers from 1749 on.

The sunspot numbers, however, are by no means as accurate an

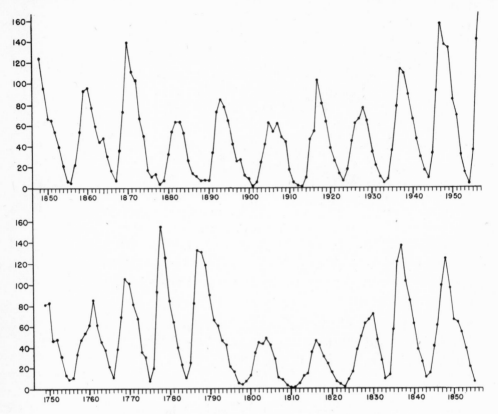

Fig. 68.   Sunspot numbers. The number for 1957 is 190.

index of solar activity as one might wish. The early estimates by Wolf from fragmentary records probably display the general periodic character correctly, but one should place little trust in the magnitude of the individual maxima.

The spot numbers ~~also~~ tend to give far too much weight to the smaller disturbances. A characteristic sunspot pair, seen as the only disturbance on the sun, would have a number of 12. Add a single small spot somewhere on the surface and the number jumps to 23, a change that amounts to almost a factor of two. The system is therefore decidedly artificial. I should recommend replacing it completely were it not for the historical value of the series. A somewhat more objective index of solar activity is the total area of all spots, umbra and penumbra, as determined at Greenwich from photographic records. In defense of spot numbers, however, we must admit

TABLE 3.   SUNSPOT NUMBERS.

Year	No.	Year	No.	Year	No.	Year	No.	Year	No.
1749	80.9	1791	66.6	1833	8.5	1875	17.1	1917	103.9
1750	83.4	1792	60.0	1834	13.2	1876	11.3	1918	80.6
1751	47.7	1793	46.9	1835	56.9	1877	12.3	1919	63.6
1752	47.8	1794	41.0	1836	121.5	1878	3.4	1920	37.6
1753	30.7	1795	21.3	1837	138.3	1879	6.0	1921	26.1
1754	12.2	1796	16.0	1838	103.2	1880	32.3	1922	14.2
1755	9.6	1797	6.4	1839	85.8	1881	54.3	1923	5.8
1756	10.2	1798	4.1	1840	63.2	1882	59.7	1924	16.7
1757	32.4	1799	6.8	1841	36.8	1883	63.7	1925	44.3
1758	47.6	1800	14.5	1842	24.2	1884	63.5	1926	63.9
1759	54.0	1801	34.0	1843	10.7	1885	52.2	1927	69.0
1760	62.9	1802	45.0	1844	15.0	1886	25.4	1928	77.8
1761	85.9	1803	43.1	1845	40.1	1887	13.1	1929	65.0
1762	61.2	1804	47.5	1846	61.5	1888	6.8	1930	35.7
1763	45.1	1805	42.2	1847	98.5	1889	6.3	1931	21.2
1764	36.4	1806	28.1	1848	124.3	1890	7.1	1932	11.1
1765	20.9	1807	10.1	1849	95.9	1891	35.6	1933	5.6
1766	11.4	1808	8.1	1850	66.5	1892	73.0	1934	8.7
1767	37.8	1809	2.5	1851	64.5	1893	84.9	1935	36.0
1768	69.8	1810	0.0	1852	54.2	1894	78.0	1936	79.7
1769	106.1	1811	1.4	1853	39.0	1895	64.0	1937	114.4
1770	100.8	1812	5.0	1854	20.6	1896	41.8	1938	109.6
1771	81.6	1813	12.2	1855	6.7	1897	26.2	1939	88.8
1772	66.5	1814	13.9	1856	4.3	1898	26.7	1940	67.8
1773	34.8	1815	35.4	1857	22.8	1899	12.1	1941	47.5
1774	30.6	1816	45.8	1858	54.8	1900	9.5	1942	30.6
1775	7.0	1817	41.1	1859	93.8	1901	2.7	1943	16.3
1776	19.8	1818	30.4	1860	95.7	1902	5.0	1944	9.6
1777	92.5	1819	23.9	1861	77.2	1903	24.4	1945	33.1
1778	154.4	1820	15.7	1862	59.1	1904	42.0	1946	92.5
1779	125.9	1821	6.6	1863	44.0	1905	63.5	1947	151.5
1780	84.8	1822	4.0	1864	47.0	1906	53.8	1948	136.2
1781	68.1	1823	1.8	1865	30.5	1907	62.0	1949	134.7
1782	38.5	1824	8.5	1866	16.3	1908	48.5	1950	83.9
1783	22.8	1825	16.6	1867	7.3	1909	43.9	1951	69.4
1784	10.2	1826	36.3	1868	37.3	1910	18.6	1952	31.5
1785	24.1	1827	49.7	1869	73.9	1911	5.7	1953	13.9
1786	82.9	1828	62.5	1870	139.1	1912	3.6	1954	4.4
1787	132.0	1829	67.0	1871	111.2	1913	1.4	1955	38.0
1788	130.9	1830	71.0	1872	101.7	1914	9.6	1956	141.7
1789	118.1	1831	47.8	1873	66.3	1915	47.4	1957	190.2
1790	89.9	1832	27.5	1874	44.7	1916	57.1	1958	184.6

that, while the relation between daily numbers and areas is poor, the agreement becomes much better in terms of the monthly and yearly averages.

So much for the general phenomenon of sunspot variability. Let us now consider more of the characteristics of individual spots.

Magnetic Fields

We have noted the occasional rapid growth or decay of a spot group. Even more surprising is the not infrequent reappearance of a spot in its original position. This behavior suggests that the disturbed areas—for there is no question but that spot regions represent some sort of upheaval—are more permanent than the individual spots.

In 1908, George Ellery Hale discovered that sunspots possess intense magnetic fields. The evidence for magnetism in spots is simple and convincing. A magnetic field will cause the ordinary atomic lines of the spectrum to split into multiple components. This phenomenon, demonstrated in the spectroscopic laboratory, is called the *Zeeman effect,* after the famous Dutch physicist who made the discovery in 1896.

The lines are not merely split under the action of a magnetic field. The individual components are *polarized;* their light waves are vibrating in certain directions relative to the magnetic field. In Chapter 3 light waves were shown to be electromagnetic phenomena, produced by vibrations of electrons. An electron may oscillate back and forth in one of an infinite variety of patterns. The simplest of these, of course, is back and forth in a straight line. If you hold a long rope in your hand and jerk it alternately up and down, a vertical wave travels away from you along the rope. The waves are said to be vertically polarized, that is, they define a vertical plane, the plane of polarization. If you move your hand and the rope horizontally, parallel to the ground, the waves along the rope display horizontal polarization. Both vertical and horizontal waves are plane polarized.

If, now, you move your hand in a circle, the rope assumes a spiral pattern. The circular motion moves away from you, along the rope. An electron, moving in similar fashion, produces what is called circularly polarized light. Or, if the vibration is an ellipse, the effect is called elliptical polarization.

The hand-rope analogy is imperfect in some respects. An electron sends out light in all directions instead of in only one. Imagine that the electron is imbedded in a huge spherical spiderweb, with each thread attached to the electron. Now, as the electron moves, the waves spread out in all directions through the web.

The vibrations are most intense in a direction perpendicular to the electron's motion. The threads that lie in the direction of motion are alternately stretched and compressed, but no waves move outward. Analogously, a vibrating electron emits no light in the direction of its motion. An electron moving in a circular path sends out circularly polarized light to points where the orbit looks like a circle, elliptically polarized light to regions where the orbit appears elliptical, and plane polarized light to where the orbit appears linear.

An individual atom usually emits polarized light. In a great mass of gas, however, the atoms are oriented at random. The light from all of these atoms thus changes its direction of polarization continuously. We say that such light is unpolarized.

A magnetic field, however, causes the atoms to line up in certain definite directions. The oscillating electrons that produce the radiation no longer vibrate at random. In a powerful field, like that existing between the poles of an electromagnet, the electrons tend either to vibrate along the field, or to swing in circular paths around the field. A spectral line produced in such a region splits in the simplest case into three components. The central component comes from the electrons vibrating parallel to the field. The two outer components arise from electrons circling the field clockwise and counterclockwise respectively.

The actual polarization and pattern depend on the direction from which one views the radiating atoms. If one bores a hole through the pole pieces of the magnet and views the luminous source *along* the field between them he will see the two outer components as circularly polarized. But the inner component will be missing because the electron is vibrating in the direction of the line of sight. If one views the light at right angles to the field, all three components will appear plane polarized, the central one parallel to the field and the two outer components perpendicular to the field.

Some spectral lines do exhibit this simple splitting into a triplet. The majority, however, display a more complex type of splitting. Each of the three main components further subdivides into a number of symmetrically spaced subcomponents. Examples of complex

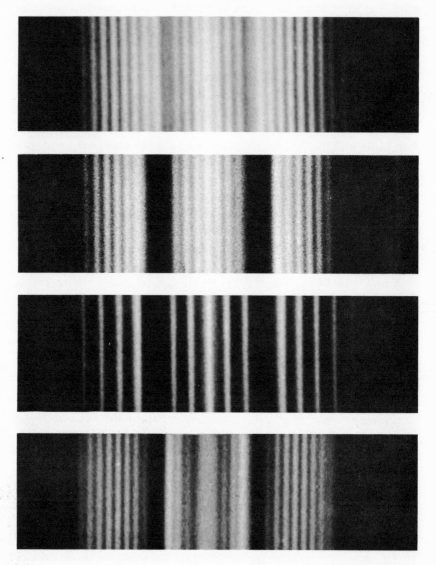

Fig. 69.   Zeeman effect in gadolinium. The Zeeman patterns of four lines are shown, each split into numerous components by the action of a magnetic field. (Spectroscopic Laboratory, Massachusetts Institute of Technology.)

Zeeman patterns appear in Figs. 69 and 70. In the former, the complete patterns show. In the latter, polarizers have separated the "parallel" or inner set of components from the "perpendicular" or outer set.

The combination of the splitting and polarization effects is unique. Only magnetic fields can produce them. When Hale found splittings and polarizations in the spectra of sunspots identical with those of the laboratory Zeeman effect, there could be no question but that intense magnetic fields existed in the spots.

A spot near the center of the disk shows two circularly polarized components. One near the limb exhibits three plane-polarized components. This observation indicates that the magnetic field of a spot tends to be perpendicular to the surface of the sun. Near the edges of spots the lines of force curve around and return toward the surface. The pattern of the magnetic field somewhat resembles that of the individual water streams from a stationary lawn sprinkler. With spectroscopic observations we can map out in detail the magnetic field of a spot region. We can even tell whether the magnetism of the spot is "north" or "south" and determine the intensity of this magnetism. Nicholson and co-workers at Mount Wilson have reg-

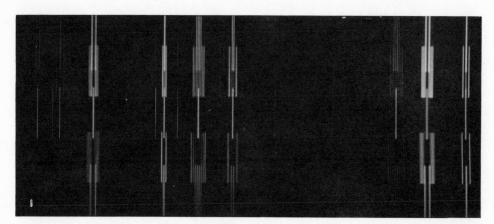

Fig. 70.   Zeeman effect in rhodium. Along the center, horizontally, we see the original lines, without the splitting. Directly above and below these are the so-called perpendicularly polarized components. At the extreme top and bottom appear the parallel polarized components. The magnetic field for the lower pair of pictures exceeded that for the upper pair, hence the greater magnetic splitting of the former. (Spectroscopic Laboratory, Massachusetts Institute of Technology.)

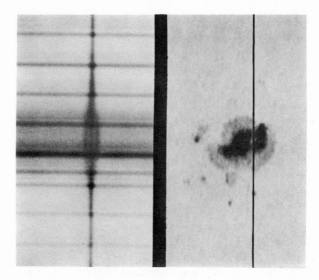

Fig. 71.   Zeeman effect in a sunspot. The position of spectrograph slit with respect to the spot is shown by the black line on the right. The vertical line on the
left is a solar absorption line. Note the split into three components in the neighborhood of the sunspot. The horizontal lines represent divisions between segments
of a special polarizing filter that will extinguish first the left-hand and next the
right-hand component of the magnetically split line of Fig. 72. (Mount Wilson
Observatory.)

ularly measured and reported the magnetic properties of all spots
since 1917.

Basically magnetic fields can arise in only one way—from electric
currents within the medium. Even in the familiar horseshoe permanent magnets, electric currents are the source. The electrons, swinging about their individual atomic nuclei, are responsible. The axes
of the tiny atomic magnets, once aligned, are held in place by the
rigidity of the steel. Merely heating the metal causes this alignment
to disappear and the magnetism vanishes. In the enormous heat of
the sun, with atoms completely free to move, there cannot be any
natural alignment of the atoms. Apparently the only acceptable
hypothesis is that spot fields arise from electric currents circulating
in the solar atmosphere on a larger-than-atomic scale.

Our studies of the solar spectrum show that many ionized atoms
exist on the sun. A free electron must accompany each ion. Thus,
if the spots were rotating vortices, would not the magnetic fields arise
naturally? The answer is an unequivocal "No"—at least not with-

out severe reservations. It is true that the rotating electrons produce a magnetic field. But a positive ion of equal electric charge accompanies each negative electron. Rotation of the positive charges also produces a magnetic field, opposite to that of the electrons. The two superposed fields just cancel. The ionization of the gas thus does not help unless there should happen to be a preponderance of electric charge of one sign or the other.

Now there is a natural tendency for electrons to escape from the sun. These particles, being so much lighter than the atoms, move

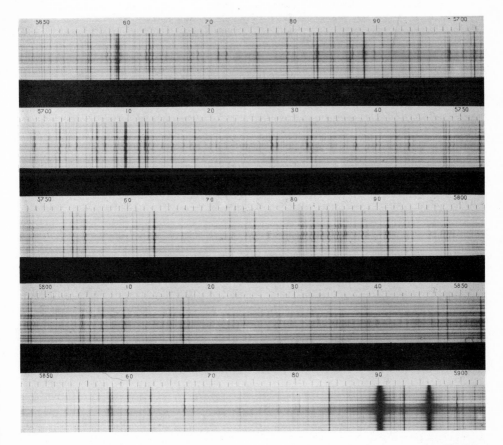

Fig. 72.   Portion of the sunspot spectrum. This picture was taken in the manner described for Fig. 71. The top and bottom of each strip represent the normal solar spectrum; the spot spectrum is along the center. The saw-tooth pattern of the lines in the spot region arises from the polarization of the light. Note that some lines are stronger in the spot than in the sun; others are weaker. The intense pair at the lower right are the well-known yellow lines of the element sodium. (Mount Wilson Observatory.)

much faster. Occasionally they may elude the gravitational pull of the sun. Hence we might expect the sun to possess a positive charge, because some of the negative electrons have slipped away. But, as Lindemann showed many years ago and Milne has confirmed, a positively charged sun exerts an electric attraction on the negative electrons, whose rate of escape is thereby limited. There can never be any tremendous separation of charge. The sun must be very nearly neutral, electrically—so nearly neutral that rotation of the gases could not possibly account for even a minute fraction of the observed magnetic fields.

Only one possibility seems to remain. The fields must arise from actual circulating currents of electricity. The electrons, because they are more mobile than the ions, slip past the other atoms and set up magnetic fields in the same way that circling electrons produce the

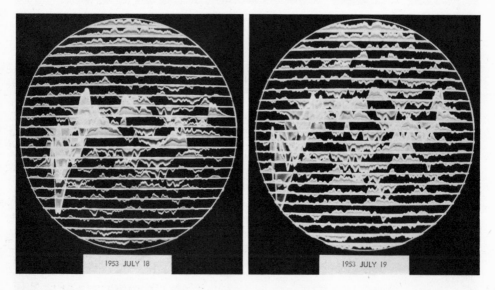

1953 JULY 18   1953 JULY 19

Fig. 73.   Solar magnetograms. These magnetic maps of the sun's disk show the location, field intensity, and polarity of weak magnetic fields in the photosphere of the sun, apart from sunspots. These records are made automatically by a scanning system that employs a polarizing analyzer, a powerful spectrograph, and a sensitive photoelectric detector for measuring the longitudinal component of the magnetic field by means of the Zeeman effect. A deflection of one trace interval corresponds to a field of about 1 gauss. The small deflection of opposite magnetic polarity near the north and south poles are indicative of the sun's "general magnetic field." The extended fields near the equator arise from characteristic "BM" (Bipolar Magnetic) regions that sometimes produce spots. North is at the top, east at the right. (Harold and Horace Babcock, Mount Wilson Observatory.)

field of an electromagnet. The one basic difference is that the currents in the sun must be in free space or at least move through a gaseous medium and not be confined to the body of a metallic conductor.

From the magnitude of the field we can calculate the intensity of the electric currents required to maintain it. A single loop of wire, 0.628 centimeter in radius, carrying a current of 1 ampere, will possess at its center a magnetic field of 1 gauss, which is the standard unit for such measurements.

The fields of large sunspots amount to as much as 2500 to 3000 gauss over areas of several hundred million square miles. We must consider a sunspot to be an enormous electromagnet. We calculate that the total current required for the field of a large spot is of the order of a million million (10^{12}) amperes.

Harold and Horace Babcock, using a magnetograph, have proved that the sun possesses a weak general magnetic field in addition to the strong spot fields we have been considering. (The magnetograph is an instrument employing a combination of optical, mechanical,

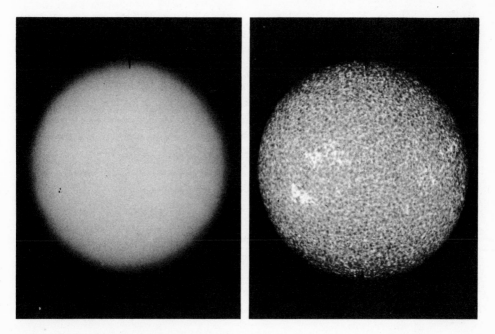

Fig. 74. The solar disk photographed on 18 July 1953 (see Fig. 73): (*left*) in white light; (*right*) in calcium light. (Mount Wilson Observatory.)

and electronic techniques for automatic measurement of the Zee-man effect.) This general field, of about 1 gauss, is somewhat weaker than the pioneer work by Hale had suggested, and it exists only in high latitudes, above $\pm 55°$. The polarity, as measured in 1955, was positive in the northern hemisphere and negative in the south-ern, just the reverse of the earth's magnetic field. During 1955 and 1956, the intensity irregularly diminished until, in 1957, the field at the south pole reversed its polarity. In early 1958 the sun possessed a positive pole in each hemisphere, but by autumn the field had reversed itself in the northern hemisphere as well. The reversal occurred 3 years after the minimum of the sunspot cycle, actually not far from the current maximum.

These observations identify the sun as a magnetic variable star. Horace Babcock had earlier discovered powerful general magnetic fields in certain types of stars. In one notable example, the field fluctuates rapidly with a period of only 9 days. I have calculated that such a star acts like a powerful radio transmitter, sending out energies of 10^{13} kilowatts or so, with a wavelength of the order of 10^{11} miles.

Babcock's magnetograph also indicates the presence of local irregularities in the general field of the sun. The most intense local fields occur in the spot zones, but appreciable deviations do appear elsewhere on the solar surface.

Although single spots often appear, records indicate, as I have previously implied, that spots tend to occur in associated pairs. Perfect examples of the double form are rather rare. The character of the associated magnetic fields, however, enables us to recognize that occurrence in pairs is a normal feature. We distinguish the two members of the group by the letters p (preceding) and f (following), according to the general direction of solar rotation.

The polarities of the magnetic fields of the individual spots of a pair are of opposite sign, as determined by spectroscopic studies of the Zeeman-effect pattern. Even when only a single spot shows, the distribution of the surrounding bright patches of calcium vapor and associated faculae often indicate the position of an additional dis-turbed region—presumably a submerged spot. In some such cases a weak magnetic field of the expected polarity has also been observed in the region.

A remarkable fact of observation, whose significance we still do not fully understand, is the opposite polarity pattern in the two hemispheres. If, for example, preceding spots in the northern hemisphere possess positive polarities, the preceding spots in the southern hemisphere will be negative. And, what is even more remarkable, the polarities of the preceding spots always shift from one cycle to the next, as discovered by Hale. Thus the real sunspot period is 22 years rather than 11, if we consider the magnetic features of the cycle.

Hale and Nicholson at Mount Wilson devised the following classification to describe the characteristics of spots and spot groups (for representative illustrations, see Fig. 88). If photographs show only one spot, with symmetric distribution of plages, the assigned classification is α. However, if the single unipolar spot lies in the preceding or following part of its plage, the designation is αp or αf, respectively.

A bipolar group, with members of approximately equal size, is denoted by the symbol β. If the preceding or following spot of the pair dominates, the designation is βp or βf, respectively. Complex spots, whose magnetic polarities are mixed, often even within a single penumbra, receive classification γ; whereas complex spots, whose major members are clearly bipolar, are symbolized by $\beta\gamma$.

From a statistical analysis of the classification of 5940 groups, observed from 1937 to 1953, Bell and Glazer obtained the following percentages of each type (21 percent of the total number being unclassified):

Classes	α	αp	αf	β	βp	βf	$\beta\gamma$	γ	Unclassified
Percentage	9	25	4	27	23	8	3	1	21

These results are in substantial agreement with earlier results by Hale and Nicholson. An examination of this tabulation discloses that "following" spots are dominant in only 12 percent of the cases. The preceding spot has a decided tendency to be larger and longer lived. The reason for this behavior is not known, but these statistical results should give an important clue to the forces that produce spot disturbances.

⊙

Some Other Properties of Sunspots

Earlier in this chapter I described a change in the average latitude of sunspots as the 11-year cycle progresses. Individual spots, however, show a remarkably small drift in latitude.

The motion in longitude is somewhat more striking. Maunder and d'Azambuja have noted that, as a spot grows in size, it tends to move slightly faster than the average. During the period of decay it slows down. This description applies particularly to the leading member of a spot pair. The trailing spot, which is usually less stable and shorter lived, has a tendency toward the opposite direction, as if the spots repelled one another.

Statistical studies, by Joy and Brunner, also reveal a tendency for the preceding spot to lie somewhat closer to the equator than its follower. This inclination of the axis of a group tends to increase with latitude, ranging from an average of 1° for equatorial groups to as much as 20° for spots in latitudes of 35°. Brunner suggests that the average inclination may decrease with time, as the spot goes through its period of development and dissolution.

Another unexplained feature is the filamentary characteristic of the penumbra, previously described. The surface of the sun, even in a region undisturbed by spot formations, is far from uniform. The studies of Janssen and others have disclosed a granular structure of bright patches on a somewhat darker background, the so-called "rice-grains." These granulations, which change their pattern completely in less than 15 minutes, result from a convective circulation in the solar atmosphere.

The suggestion has been made that the penumbral filaments consist of distorted granules, drawn out into long wisps by atmospheric circulation. Or perhaps they represent part of a horizontal or slightly inclined convection stream, which would show as a granule outside the spot. The effect is similar to that of a bundle of matches, which shows the heads or the sticks according to the angle of view. We may assume that the "sticks" show up on the inclined edges of spots, which slope downward to the dark umbral core.

Strong magnetic fields completely inhibit convection within the umbra. Outside the spot, convection is dominant. In the penumbra, the opposing forces of convection and magnetism are probably nearly equal and we may suppose that here the convective granules are stretched out and oriented by the magnetic field.

Spots show a wide variety of deviation from the simple, circular form. The umbra may be irregular, stellated, or compressed. The penumbra tends to follow the umbral contour, at least roughly, but the extent of the penumbra may vary considerably with spot size. Pores or small spots possess little if any penumbra. A large complex group will sometimes have a single extensive penumbral region enveloping several umbras. Both the inner and outer edges of a penumbra are usually sharp, despite the irregularity of contour.

There is some evidence to indicate that a spot is a depression in the solar surface. In 1760, Wilson of Glasgow noted, near the limb, a spot that showed more penumbra on the side farther from the center of the disk. If all spots were absolutely symmetrical, this observation would definitely prove that spots are hollows. The irregularity of penumbral rims makes the conclusion less definite. If we can believe the statistics, however, the penumbral regions are gentle slopes, with the umbra lying 500 miles or so below the level of the outer rim. There is some possibility, however, that the outer edge of the penumbra is somewhat elevated, so that one cannot speak with certainty of the relative altitudes of the umbra and the normal solar surface.

Gaseous circulation in the neighborhood of a spot is very complicated. Evershed, working at Kodaikanal Observatory in India, was the first person to determine the general characteristics of the motions of these gases. Setting the slit of his spectrograph across the

Fig. 75.   Large sunspot group of 17 May 1951. Note the penumbral filaments and photospheric granulation. (Mount Wilson Observatory.)

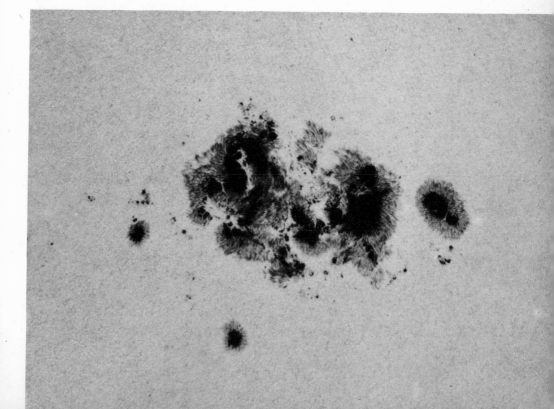

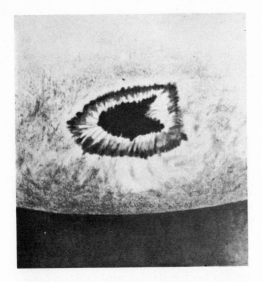

Fig. 76. Drawing showing a sunspot as a depression. (Rordame.)

spot, he observed the peculiar and distinctive distortions of the lines, indicative of radial velocities.

When the spot was near the center of the disk, the line displacements were usually negligible. This result indicated that the gases showed little motion toward or away from the observer, that is, up and down in the spot itself.

But when the spot lay at some distance from the center of the disk and when his spectrograph slit coincided with the solar radius, Evershed found marked displacements in the penumbral regions. These shifts indicated motions toward the observer on the side nearest the center of the disk and away from the observer on the opposite side. The gases in the spot were moving radially away from the spot center, parallel to the solar surface. This phenomenon we generally call the *Evershed effect*.

The foregoing description holds for spectral lines of low and medium intensity. The strongest lines of all, those of ionized calcium and neutral hydrogen, appear to show an opposite shift, which we interpret as motion inward. Now, the calcium and hydrogen lines originate at somewhat higher atmospheric levels than do the other lines. These observations suggested that a sort of flattened, smoke-ring vortex lies above the spot, with outflowing gases in its lower part and inflowing gases in its upper regions. Evershed found, however, no trace of vortex motion in the horizontal plane.

A somewhat more recent study of the phenomenon, by Abetti in

Arcetri, Italy, indicates that the circulation may be even more complex. Abetti finds that the motions described by Evershed are only statistically correct. The velocities differ markedly from spot to spot. Also, whereas Evershed found no definite indication of cyclonic rotation of the spot, Abetti found evidence for a small component ranging from 1 to 3 kilometers per second.

The observations of motions of gases in spot regions are only part of the general problem of solar atmospheric circulation. St. John and his Mount Wilson collaborators have found a small velocity of ascent for the faint lines produced at very low photospheric levels. Gases producing the stronger lines, on the other hand, seem to be moving downward. At the limb of the sun, these shifts caused by vertical circulation vanish.

The interpretation of line displacements however, is complicated by the superposition of several effects, such as variable circulation, unsymmetrical broadening of lines by collision with other atoms, and the so-called "Einstein effect." The last, a minute shift toward the red of spectral lines produced in a gravitational field, appears to agree with the theoretical value.

Many phenomena associated with spots deserve further study, and I shall discuss some of them in later chapters. Here, however, it is appropriate to consider the faculae, the bright, veined networks around the spots. A few days or hours before a sunspot becomes visible—perhaps while it is forming beneath the surface—a cluster of bright clumps and veins may appear in the region. Then suddenly, in the course of a few hours, a hole may appear in the facular patches and grow into the umbra and penumbra of a sunspot. In other cases the development of spots and faculae appears to occur simultaneously. More consistently, the faculae will remain to mark the disturbed area long after the spot itself has died away. No spot of any size is without its surrounding faculae, although the latter are clearly visible only near the limb of the sun. But faculae often occur without any visible spot.

The faculae are most conspicuous in the spot zones, but in more fragmentary form they are visible in all solar latitudes. It is possible that the small isolated polar faculae may arise from some turbulent activity of a nature entirely different from that occurring in the spot regions. But their positions may indicate disturbed regions whose activity is not sufficient to break through the visible surface and form a spot.

In 1875 Trouvelot called attention to a solar phenomenon that has since received but little notice. He noted occasional grayish patches, which he termed "veiled spots." Unlike ordinary spots, they sometimes appeared within 10° of the poles. I have myself seen hazy markings—blotches, patches, or filaments—that do not fall into any standard classification of solar features. They show up in white light, without the aid of special filters. W. O. Roberts has occasionally noted patches that resemble penumbral regions, but without the darker umbra. They look like buried spots and may be identical with the phenomena described by Trouvelot.

The bright calcium clouds, called flocculi or plages, which are associated with the faculae, are equally sensitive to the presence of spots, and often strikingly delineate the position of the spot zones in both hemispheres. For details, see the next chapter.

Theories of Sunspots

We have learned that sunspots are dark and cool, that they possess powerful magnetic fields, and that their number varies characteristically with a period of 11 years. Analogy with the earth's atmosphere, where storms are cold areas of low pressure, long ago suggested to astronomers that sunspots might be storms. In the first edition of this book, I titled this chapter, "Sunspots—Solar Cyclones." Recent work has proved that this once popular picture of a spot as a storm is completely wrong.

The motions of the earth's atmosphere consist of two basic varieties—vertical and horizontal. To consider vertical flow, imagine that I hold in my hand a large balloon, filled with air and covered by a weightless, elastic skin. If we further suppose that the temperature and pressure inside equal those on the outside, the balloon will neither rise nor fall when I release it.

Now suppose that I raise the balloon over my head, into a region of lower air pressure. The balloon expands and the air inside cools with the expansion. Now measure the temperature inside the balloon and compare it with the outside temperature at the new height. If the balloon is cooler than its new surroundings, the gas inside will be denser, and therefore heavier, than the air outside. The balloon, in consequence, will fall and tend to return to its initial position. But if the interior should be hotter, the gas will be lighter and the balloon will tend to rise. If the balloon should have an internal source of heat, it would rise even faster.

Even though the earth's atmosphere is not encased in weightless, elastic balloons, great masses of air are subject in an analogous way to forces caused by irregularities of temperature. In consequence, hot air will rise and cold air descend. As with the balloon, the lifting forces will be greater if each ascending mass possesses an internal heat source. Moist, warm air does provide for extra heating; when the temperature of the rising, expanding gas decreases to the point where water vapor starts to condense, the process of condensation tends to heat the gas. For this reason, the billowing cumulus clouds are often violently turbulent. Airplane passengers will complain of "bumpy" air. This convection may even be dangerous to smaller planes.

In the sun's atmosphere energy is transported by a combination of two processes: convection and radiation. We have recently found that convection is more important than we had previously realized. Unsöld pointed out that hydrogen, which is ionized at low levels and neutral at higher levels in the sun's atmosphere, performs the same service for the sun that water vapor does for the earth's atmosphere. In a rising, expanding, and therefore cooling mass of gas, the ionized hydrogen tends to become neutral. The recombination of electrons and protons releases energy, which goes into heating of the gas. In consequence, the outer solar layers should be violently convective.

So much for the vertical circulation. On earth, the horizontal circulation depends primarily on the flow of cold air from the vast polar ice caps, where the air becomes denser and builds up a region of high pressure that spills over toward the equator. The hot, equatorial air rises and flows poleward in an attempt to equalize the pressure. The circulation, however, is by no means smooth. Areas of high and low pressure develop in intermediate latitudes. These latter are the centers of stormy weather.

Air flows from the high- to the low-pressure areas, expanding and cooling in the process. Rotation of the earth makes an important contribution to the action. The polar air masses have a lower velocity of rotation than the atmosphere in intermediate latitudes. Consequently, as this air flows southward from the north pole, it tends to lag behind. The air stream deviates to the west. Conversely, as the rapidly rotating equatorial air blows into higher latitudes, the deflection is to the east. The air caught between these streams thus tends to rotate and a whirlpool motion or cyclone develops.

The general circulation of winds about a terrestrial storm center, as seen from above, is thus always counterclockwise in the northern hemisphere and clockwise in the southern. The speed of rotation increases toward the center. Hence, if the storm is extremely violent, the whirling gases produce a hurricane.

There are, of course, many complexities of the earth's atmospheric circulation not covered by the foregoing brief survey. One may mention the so-called "fronts" where cold, denser air may "snow-plow" under a lighter mass of warm air. Also vertical instability plays a very important role in the formation of tornadoes and hurricanes. Friction of the land must be considered, together with a host of other factors, such as moisture content—all of which make weather prediction a difficult science indeed.

These facts of the horizontal terrestrial circulation led to the conventional view of sunspots as vortical storms resembling hurricanes. In an attempt to account for the peculiarities of spot development, both Hale and Bjerknes developed theories of vortex circulation. The opposing magnetic fields of the members of bipolar groups suggest that the rotation should be opposite for the two spots.

If the preceding spot is rotating clockwise, for example, mere viscous friction would tend to make the following spot move in a counterclockwise fashion. Hale suggested that the two vortices were connected at the base. To picture his model, imagine that you hold a two-foot length of flexible garden hose, bent into U-shaped form with an end in either hand. Now, if you rotate the right-hand end clockwise, the physical connection will cause the left-hand end to rotate in the opposite direction. Thus, Hale proposes that the bipolar groups are opposite ends of a U-shaped vortex, as in Fig. 77.

Bjerknes also assumes that the spots are coupled, but he suggests that the U, instead of dipping only a short distance below the surface, actually encircles the sun along a parallel of latitude. This vortex, Bjerknes further postulates, is essentially permanent. It is buried below the surface, twisting and turning like a gigantic sea serpent, until a disturbance cuts the tube. Then one or both of the severed ends may rise to the surface and form a group of spots until forces of an unknown character repair the break (Fig. 78).

Bjerknes further suggests that there are four great doughnut-shaped vortices, two in each hemisphere. In the northern hemi-

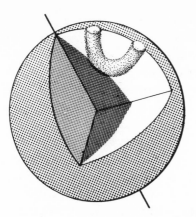

Fig. 77.   Sunspot according to Hale.

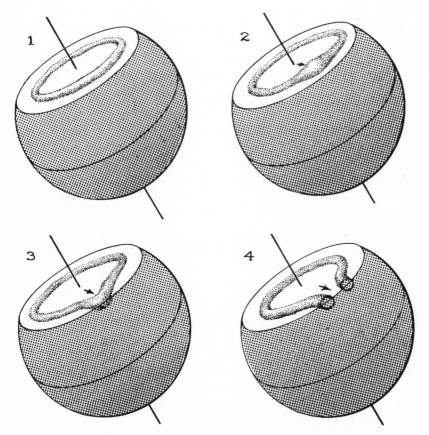

Fig. 78.   Theory of Bjerknes.

sphere, we are to visualize the vortices as a pair of smoke rings, one within the other. The outer and larger vortex, which produces the visible spots, slowly drifts toward the equator, while the inner and smaller vortex moves poleward. The first vortex shrinks and sub-merges as it reaches the equator; meanwhile the second one expands and rises. Now the roles are reversed and the second vortex moves equatorward. But, since the direction of circulation is different for the two vortices, the magnetic polarities appear reversed. Analogous activity takes place simultaneously in the southern hemisphere.

These theories are attractive, but they completely ignore the most significant feature of spots—their magnetic fields. Alfvén, of Sweden, has suggested an alternative hypothesis. He assumes that magnetism is present in the solar core. If now, near one edge of this inner sphere, a sudden disturbance occurs—an explosion or eruption—a wave or pulse will spread out and slowly move to the surface. This wave will carry with it, according to the theory, some of the intense magnetic field of the interior.

Walén, also of Sweden, has developed the suggestion in a more definite form. He indicates that the pulses are doughnut-shaped vortex rings. These rise very slowly to the surface. A bipolar spot results when the visible exterior takes, in effect, a bite out of the doughnut. The portion remaining below the surface thus resembles the coupled-hose picture originally proposed by Hale.

Alfvén and Walén identify the 11-year cycle as the time required for the pulse to travel from one side to the other of the core, which was then thought to be convective. A disturbance, starting from the edge of the region, will form—according to the hypothesis—two rings. The one moving outward will behave as explained in the previous paragraph. The other will first move into the center and will again reach the edge of the core only after the lapse of 11 years.

The picture presented by Alfvén and Walén is qualitatively appealing, but much more work needs to be done before the hypo-thesis can be accepted. Neither theory accounts satisfactorily for the reversal of magnetic polarity during the latter half of the 22-year cycle. Walén subsequently rejected his original theory in favor of a model in which the rate of the sun's rotation is variable. Cowling has raised some objections, especially with respect to the magnetic phenomena described.

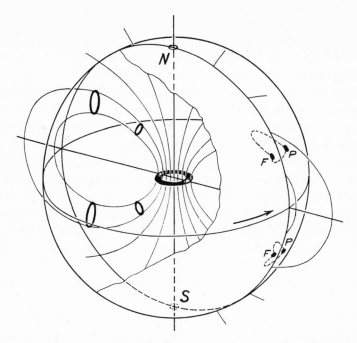

Fig. 79.   Solar model of Walén. Note the doughnut-shaped vortices shot from interior and reaching surface as spots.

The vertical convection appears to account satisfactorily for the phenomena of granulation, faculae, and flocculi. We are to visualize the solar surface as violently turbulent, covered with enormous waves several hundred miles across. As for sunspots, it appears that we have stretched too far the analogy between the solar and terrestrial atmospheres. In the latter, horizontal circulation results from the very unequal heating between pole and equator. On the sun, no such major temperature differences exist, so that north-south circulation cannot be particularly violent. Sunspots, therefore, cannot arise from horizontal circulation.

The intense magnetic fields of spots provide the major clue. As Biermann pointed out, magnetism tends to bind together neighboring masses of gas, inhibiting flow of gases across magnetic lines of force. A rising gas cannot, therefore, flow horizontally and descend again to complete the convection cycle. As a result, spots are calm regions in an otherwise turbulent atmosphere. Since convection is absent, energy flow is restricted to radiative processes, which are less efficient. Consequently, as Krook, Wild, and Menzel have suggested, sunspots are cool because convective energy transport is less in the presence of magnetism. And, in the regions just outside the spots, convection must be greater to make up for the lessened heat flow over the spot areas. The bright facular and floccular patches result from heightened convection in the neighborhood of spots. These areas are brighter because convective flow has brought more of the hotter gases nearer the surface.

We still do not know precisely what causes sunspots. Krook, Layzer, and Menzel have carried out a semiquantitative analysis. The reversal of polarities between successive cycles suggests that the general magnetic field below the surface may be stretched out, as if a giant comb had been drawn around, parallel to the equator. For one cycle, the lines are drawn eastward, in the direction of solar rotation; in the next cycle, the lines of force are drawn in the opposite direction. Some sort of oscillating, torsional vibration in

Fig. 80.   Curve of a single sunspot cycle, with four calcium spectroheliograms taken at different stages. The bright calcium clouds, which generally appear in disturbed areas, are far from the equator at the beginning of the cycle. Thereafter they tend to move nearer the solar equator. (Nicholson, Mount Wilson Observatory.)

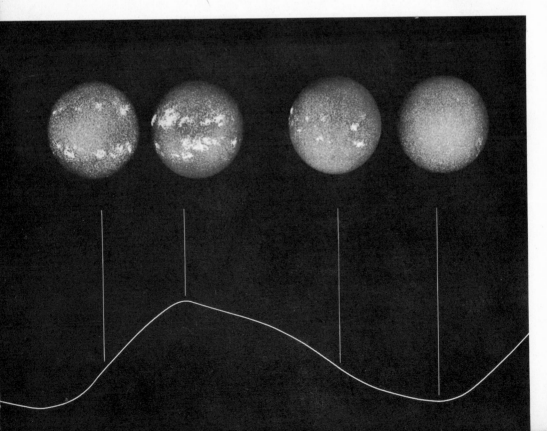

the sun appears to be responsible for this behavior. But this model has some difficulty in accounting for the reversal of the general field.

Subsurface convection, carrying the material outward, stretches the lines of force, increases the magnetism, and accounts at least qualitatively for the major features of solar activity. The hydrogen convective zone, which extends to a depth of perhaps one-tenth of a solar radius, is largely responsible for carrying this material. The fact that gases containing magnetic fields possess a balloonlike buoyancy, as Parker has noted, tends to help spot formation. On this view, strands of magnetism—some perhaps shaped like dough-nuts—disconnect themselves from the general field and float to the surface. The 22-year periodicity measures the time of one complete oscillatory motion of the sun. This theory is incomplete, but it accounts far better for the characteristics of spots than do the older ones, which depended on the presence of vortices and storms.

$\odot$

Observations for Amateur and Professional

Sunspot and solar studies are fruitful fields for amateur astronomers. Members of the solar division of the American Association of Variable Star Observers daily estimate the sunspot numbers, note the locations of disturbed regions, mark facular areas, look for the presence of fine granulation, and so on. Some of the amateurs have fitted their telescopes with cameras and send daily pictures to Zurich, Switzerland, for permanent filing.

Smallness of the telescope is no bar to useful work. The observer may project the sun upon a piece of white paper, or employ various kinds of glass filters to reduce the brilliance to a level suitable for eye comfort. Students possessing *neutral filters,* which do not alter the color values, sometimes have reported seeing a rosy or brownish light over some of the spots. This coloration probably arises from prominence activity in the region and, in looking for this phenom-enon, one must take special care to eliminate the artificial colors that occasionally result from telescopic imperfections.

Solar observation brings satisfaction to professional and amateur astronomer alike. The ever-changing solar panorama, as spots develop and dissolve, as they slowly rotate with the sun, kindles the imagination. One always looks forward to seeing what changes have taken place during the night. Large and active areas suggest that the earth may be disturbed, exhibiting magnetic storms, brilliant auroras, radio disturbances, or variability of cosmic radiation. And when such terrestrial phenomena do occur, observers realize how intimate is the relation between sun and earth.

There are still many unsolved problems related to sunspots. Some seem so complex as to defy solution. The one consoling and encouraging fact is that we have already advanced far. During the first half of the nineteenth century, the generally accepted hypothesis of sunspot constitution was that proposed by Sir William Herschel. The spots were supposed to be holes in the sun's fiery envelope through which we could glimpse the cool and presumably habitable solar surface. The concept of sunspots as cyclones or solar storms that replaced the earlier view has now proved to be equally untenable. How easy it is for our preconceptions to dominate our conclusions!

7

Fine Details of the
Solar Surface

Solar photography, with the aid of special color filters, has revealed remarkable surface details. The familiar color picture, photograph or lithograph, derives its color effect from the superposition of at least three pictures, taken in the light of different colors. For example, three cameras may record simultaneous views through glass filters: red, yellow, and blue. The combination of the three separate positives by any one of a dozen different methods produces a result whose color values approach those of the original.

From the scientific standpoint the eye is a poor judge of color. Experiments have shown that the same color sensation may result from a wide assortment of different shades. The physicist and astronomer require a definition of color more precise than that given by subjective impression. The accurate measure is in terms of wavelength, as discussed in Chapter 3. Moreover, the concept of wavelength applies to invisible ultraviolet or infrared, as well as to visible light. The spectroscope or spectrograph is the tool by whose means we determine the color and quantity of the radiation.

Every camera fan is familiar with ordinary glass filters. He knows that a yellow (K2 or K3) filter will remove some of the blue light of the sky and produce enhanced cloud effects in scenic pictures. A red (A) filter may so darken the sky that the picture will appear to have been taken in moonlight. The red filter may also remove the effect of haze and clearly show a distant mountain range.

By placing one of these glass filters in the front of a spectrograph we can see just how much light gets through in each color. One fact emerges immediately from such an experiment. The bands of color transmitted by the glass, however pure they may seem to the eye, are extremely broad. The K2, yellow, filter transmits a tinge of green and all of the yellow, orange, and red.

Although such wide-band glass filters have their astronomical uses, a narrow-band filter, which passes only a minute color range, would be more valuable for solar study. We speak of such a filter as *monochromatic*. We have seen that the solar spectrum displays a number of dark lines, each arising from the presence of some chemical element in the solar atmosphere. If our filter could pass the light from only one of these narrow lines, we should be able to find out how the shining atoms of that substance are distributed over the solar surface.

To reach this narrow-band goal we need a filter a thousand times or so more efficient than glass. The resolution required approaches that of the spectroscope itself. In fact, why not employ the spectroscope for this very purpose?

In 1868, Janssen, observing a total eclipse in India, was struck with the brilliance of several bright lines from the solar prominences at the limb. As soon as the eclipse was over he turned his instrument on the sun again and was delighted to find that he could still see, in the red light of hydrogen, the same bright patch he had noted during the eclipse. By opening up the slit of the spectroscope, he could even discern the form of the luminous gas. Sir Norman Lockyer, in England, made the same discovery independently and simultaneously.

For many years, this method of observing the prominences was the only one available to astronomers. It had the decided limitation that the procedure was useful only for prominences projecting

Fig. 81. The 150-foot tower telescope at Mount Wilson, which feeds a spectroheliograph located in a deep well beneath the tower. The telescope is used for solar studies. (Mount Wilson Observatory.)

beyond the sun's limb. It gave no hint of phenomena occurring on the solar disk. Young, of Princeton, one of the outstanding pioneers in solar research, employed the open-slit spectroscope for many years in his study of the sun.

⊙

Spectro-
heliographs   Janssen had suggested a possible adaptation of the spectrograph for photographs in a single color. We have already noted that the lines of the spectrum are, in reality, images of the spectrograph slit. If the illumination on the slit is nonuniform, for any reason, the line image will show the same pattern.

Let us permit the solar image to drift across the spectrograph slit. The details of the solar disk will be successively depicted in the distribution of intensity along the spectral lines. Now place a second slit in the position of a line we wish to study, to isolate a monochromatic region of the spectrum. Finally, if we cause a photographic plate to move across the second slit at a rate synchronized with that of the solar drift, we build up on the plate a picture of the entire sun in the light of the chosen color. In effect, we are cutting a picture into a thousand or more linelike strips and then reassembling them into the original form, as does an ordinary television set.

Hale and Deslandres, practically simultaneously, built instruments on this principle. We call the device the spectroheliograph— a compound word signifying photography of the sun by means of the spectrograph.

⊙

Spectro-
heliograms   The results obtained with such an instrument proved to be little short of sensational. The images of the solar disk, depicted in the light of hydrogen or calcium, disclosed phenomena whose existence had never been suspected. The spectroheliograms were entirely unlike direct photographs. True, sunspots sometimes showed up as dark holes, but the entire disk was marked by splotches, filaments, and streamers, presumably of the luminous gas whose light had passed through the second slit.

Spectroheliograms in the light of ionized calcium show character-

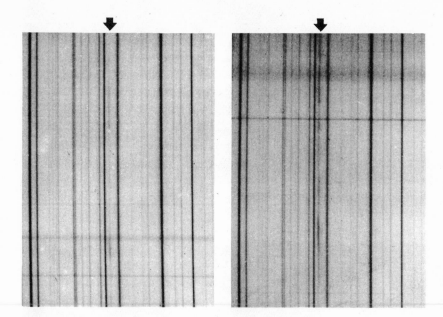

Fig. 82.   Two portions of infrared spectra of the solar disk. Note that the hard-to-excite helium line, indicated by the arrows, varies in intensity over the solar disk. (d'Azambuja, Meudon.)

istic forms that differ appreciably from those taken in the red line of hydrogen. The former display, as their most distinctive feature, large bright patches of calcium vapor in the vicinity of spots or disturbed areas. We refer to these clouds as calcium *flocculi* or *plages*. The hydrogen pictures show much finer details. The bright flocculi are smaller and often show an irregular filamentary structure. The background is speckled in contrast to the coarser mottling seen in calcium light.

The differences between the calcium and hydrogen photographs certainly cannot represent actual variations in chemical composition. The solar atmosphere is far too turbulent. Its constituents must be well mixed. We must attribute the effect rather to some characteristic of the chemical elements in a nonuniform environment.

I mentioned in Chapter 5 that most of the calcium atoms in the solar atmosphere are able to produce the violet lines. Only one hydrogen atom in a million, however, can produce the red line, Hα, under average solar conditions. Hydrogen is particularly difficult to excite. A slight increase in temperature will put calcium into emission, whereas Hα needs a much greater increase.

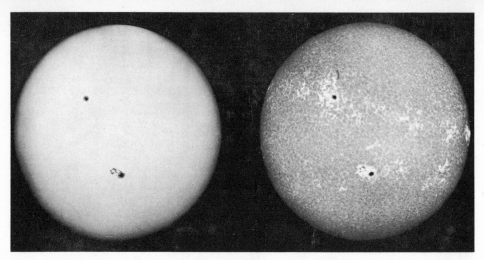

Fig. 83. The sun, direct image and calcium (K) spectroheliogram, 30 July 1906. (Mount Wilson Observatory.)

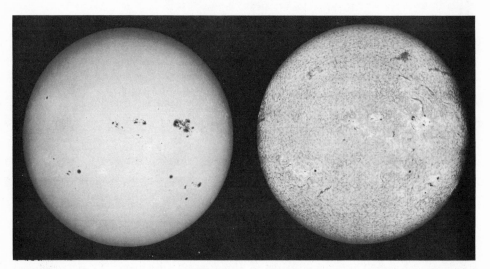

Fig. 84. The sun, direct image and hydrogen (Hα) spectroheliogram, 12 August 1917. (Mount Wilson Observatory.)

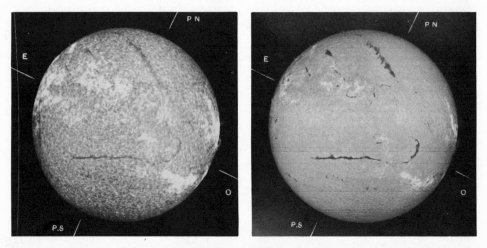

Fig. 85. Solar spectroheliograms, 25 March 1949: (*left*) calcium (K); (*right*) hydrogen (Hα). (d'Azambuja, Meudon.)

A patch of bright hydrogen vapor therefore indicates that the region must be very hot. And, since cool hydrogen gas is nearly transparent, a dark patch of absorption means that some excitation, even though less extreme, is still present. We do not completely understand why some hot patches appear bright and others dark. Probably, however, the brilliant regions are much hotter, denser, and thicker than the darker areas.

Long, dark, irregular clouds often appear on both calcium and hydrogen spectroheliograms. These dark patches, called *dark flocculi* or *filaments,* generally possess a fibrous structure. On small-scale pictures, they may suggest long black worms against the surface (Fig. 85). But high magnification shows that they are far from solid. Hundreds of fine threads closely interlaced make up the body. These formations are relatively long-lived. One of them may persist for days, only to break up in a few hours of violent activity.

Dodson and McMath noted the sudden vanishing of such a filament that had persisted for several weeks. Surprisingly, this filament appeared again within a few days, in substantially its old position on the disk.

We know that these long, dark flocculi are really prominences, enormous clouds of glowing gas that often extend to great heights above the solar surface. Seen in projection against the brilliant solar disk they look black; but they appear luminous against the dark sky beyond the edge of the sun.

The bright hydrogen flocculi, by contrast with the dark, are usually associated with sunspots. These flocculi are by no means structureless. Each patch is reticulated, broken up into smaller areas outlined with veins somewhat brighter than the background. The veins themselves may not be continuous. Knots lie along them at irregular intervals, like beads on a string. These bright patches are restless objects, constantly changing in detail. Individual features may last for several hours or they may alter strikingly in a matter of minutes.

The fine-grained structure of the hydrogen pictures (Fig. 87), in the neighborhood of spots, often displays a distinctive pattern suggestive of that taken by iron filings in the vicinity of a magnet. One finds it difficult to escape the conclusion that the magnetic fields of the spots have something to do with this formation. The brightest (and presumably hottest) areas usually lie between the main spots of the bipolar group. Neighboring dark filaments usually show cur-

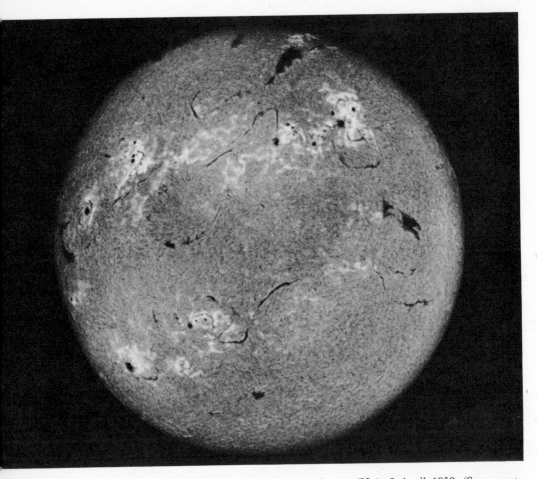

Fig. 86.   Solar disk in the light of hydrogen (Hα), 6 April 1958. (Sacramento Peak Observatory.)

vatures that conform roughly to the general pattern of the magnetic lines of force. Note also, in Fig. 87, the structureless character of the penumbral regions.

Striking increases in the brightness of the luminous gas occasionally occur. These *flares* (Figs. 98 and 99) are best seen in hydrogen light, though they are frequently visible in other radiations and occasionally show up even in white light. A flare is generally a brightening of a small part of an existing plage structure and shows a similar veined or beaded appearance. Flares commonly occur in the patches of bright hydrogen flocculi surrounding sunspots.

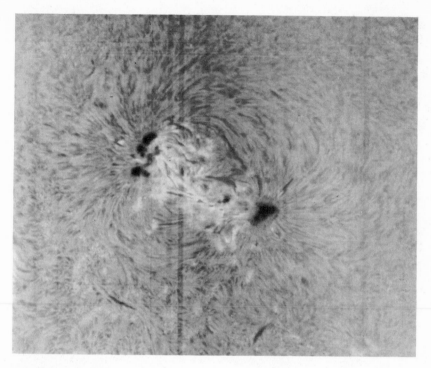

Fig. 87.   Hydrogen spectroheliogram of a bipolar sunspot group. (Mount Wilson Observatory.)

The speckled, salt-and-pepper background of minute bright and dark hydrogen areas may change rapidly, often in 10 or 15 minutes. Motion pictures of the surface present a sort of "crawly" appearance—like white worms in a pile of carrion. A somewhat less repulsive metaphor would compare the changing solar scene to white-caps dancing madly in a stormy sea. The phenomenon undoubtedly represents an effect of violent convective flow.

Spectroheliograms taken in calcium light show some features in common with those of the hydrogen photographs. Both exhibit the dark flocculi in almost identical detail. Bright eruptions and flares are most common in the disturbed regions around the spots, but the plages are almost uniformly bright in the calcium pictures, and lack the detailed structure so often seen in hydrogen. In calcium a general large-scale mottling replaces the speckled hydrogen background.

Nicholson employs the general appearance of calcium spectro-

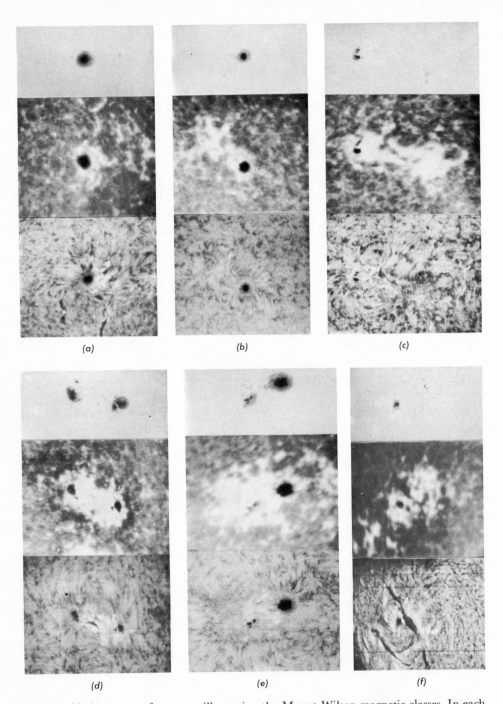

Fig. 88.   Various types of sunspots, illustrating the Mount Wilson magnetic classes. In each figure the top photograph is in ordinary light, the center picture is in the K-line of ionized calcium, and the lowest is in the light of hydrogen (Hα): (*a*) a unipolar spot (α); (*b*) a unipolar spot (α*p*) that precedes (lies to the west of) a bright calcium plage, which indicates a disturbed region even if no true spot is visible; (*c*) a unipolar spot (α*f*), following a bright flocculus; (*d*) a bipolar spot (β); (*e*) a bipolar spot (β*p*), the preceding spot of the pair being dominant; (*f*) a bipolar spot (β*f*), the following member being dominant; (*g*) a complex bipolar spot (βγ); (*h*) a multipolar spot (γ). (Mount Wilson Observatory.)

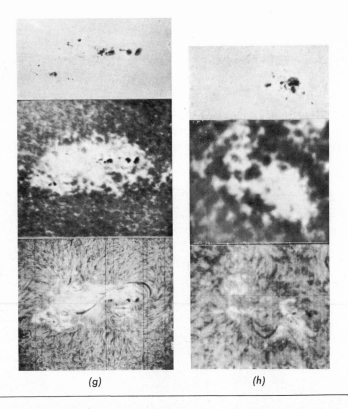

(g)   (h)

heliograms as a part of the basis for his classification of sunspots
(p. 117). Figure 88 shows characteristic examples of the various
types of spots. The upper record in each picture is taken in unfiltered light; the others are, respectively, records in calcium (K) and
hydrogen (Hα) light.

Why the very bright calcium plages around sunspots? Although
the spot itself is a quiet region, held relatively motionless by its
magnetic field, the immediate surroundings of the spot are more
active and turbulent than elsewhere on the solar surface. Throughout the disturbed region, the atmosphere is turbulent and more
distended. The gases are hotter and prominences more active than
in neighboring areas. These factors combine in general to increase
the emission from the disturbed locality.

Some of the differences between hydrogen and calcium photographs may be attributed to the fact that the absorption line of calcium is about ten times broader than that of hydrogen. The calcium
plages come from the center of the line. To take a similar picture
of hydrogen would require a filter narrower than any yet available.

There remain, however, distinctive differences. Compare Fig. 89a (K of calcium) with Fig. 89b (Hα of hydrogen). Although the bright hydrogen flocculi appear in the same region as the calcium clouds, they are generally much less extensive. Most of the fine details match well, as inspection will confirm, but the dark flocculi are relatively more conspicuous on the hydrogen pictures.

Most spectroheliograms are taken in light of the K line of ionized calcium or in the Hα line of hydrogen. The other lines of the solar

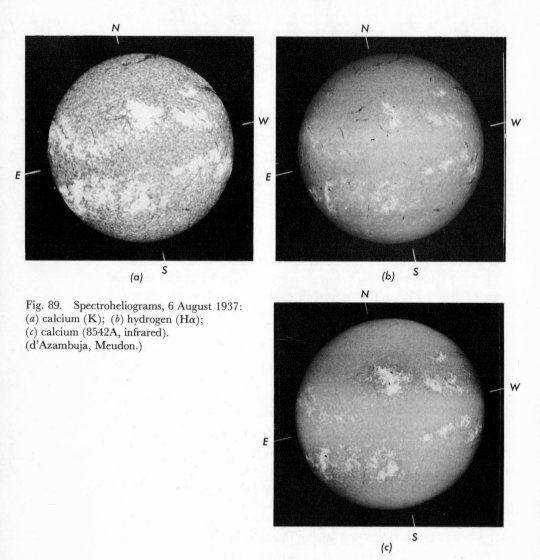

Fig. 89.   Spectroheliograms, 6 August 1937:
(a) calcium (K); (b) hydrogen (Hα);
(c) calcium (8542A, infrared).
(d'Azambuja, Meudon.)

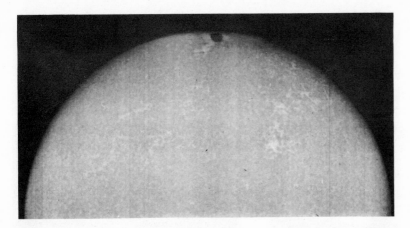

Fig. 90.   Spectroheliogram in neutral calcium, 4227A. (d'Azambuja, Meudon.)

spectrum are too faint and narrow for study under ordinary condi-tions. However, M. and Mme. d'Azambuja, of Meudon, have used special techniques and remarkable skill to obtain images in the light of several additional lines.

A record in the 8542 A line of ionized calcium, a line in the infrared, much fainter than the K line, appears in Fig. 89c. Com-pare this picture with Fig. 89a, for they were taken simultaneously. The mottled appearance of the K-record is largely gone. The bright patches are less extensive, but the dark flocculi more nearly resemble those of the other calcium record than the hydrogen. Figure 90 shows a record obtained in light of neutral calcium (4227 A). The great plages show up as mere veined streaks, scarcely more con-spicuous than faculae.

Figure 91a shows a small active area where the yellow line of helium appears in emission near a spot. As previously stated, helium is an atom that requires a particularly high temperature (25,000° K) for its appearance, in either absorption or emission. The association of the bright helium with the calcium plages clearly indicates that these areas are the hottest regions of the solar atmosphere. Helium shows up in emission only in the most intense solar flares. This yellow line rarely appears in absorption, but the stronger related infrared line of helium, at 10,830 A, does show the dark filaments.

The three illustrations of Fig. 91b, c, d are remarkable. The first two are generally consistent with the appearance of calcium and

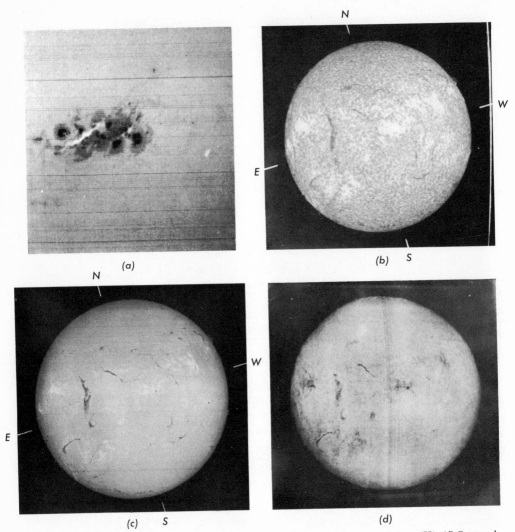

Fig. 91.   Spectroheliograms: (*a*) helium (5875A), 25 July 1946; (*b*) calcium (K), 17 September 1938; (*c*) hydrogen (Hα), 17 September 1938; (*d*) helium (infrared, 10,830A), 17 September 1938. (d'Azambuja, Meudon.)

hydrogen spectroheliograms. The last, however, is a record taken with the infrared helium line at 10,830 A. Note that emission is negligible, and areas bright in the hydrogen picture show dark in the helium photograph.

⊙

I have referred earlier to motion-picture photographs of solar activity. These records mark one of the most significant advances to date in the understanding of solar phenomena. Three enthusiastic amateurs—the late Francis McMath, his son Robert R. McMath, and their lifelong friend, Judge Henry S. Hulbert—built a solar observatory equipped with mechanical and optical equipment of highest precision. The boldness of their concept, the skill of their engineering, and the high quality of their scientific results have placed this observatory among the leaders in solar research.

Motion-Picture Recording

Robert McMath has received world-wide recognition for his fundamental studies. He and his colleagues were pioneers in the new field of *spectroheliokinematography*. Break this word into its components and we have "spectroheliograph" and "motion pictures" (kinema). These men used the basic principles for solar study as developed by Hale in the spectroheliograph. After greatly improving the speed of the instrument, they added motion-picture recording.

A single picture requires less than 30 seconds for its recording. The time of exposure is adjustable. The successive "stills" are all of the same section of the sun. Projected at the standard rate, from 16 to 24 frames per second, the pictures are speeded up by 300 to 600 times over the original solar rate.

The eye cannot easily follow the slow changes on the sun. Even the successive pictures taken with the ordinary spectroheliograph— one every 15 minutes or so—failed to give the impression of continuity obtained by motion-picture recording. The McMath-Hulbert photographs gave the scientist, for the first time, a dynamic presentation of solar activity.

Some of the most spectacular results obtained with this instrument are those dealing with rapid changes in the variable bright and dark flocculae. These consist of plages, flares, and filaments seen in projection against the disk. In an active region, the luminosity often appears to flash along the filaments from one end to the other, giving an impression of pulsating activity.

Occasional very brilliant flashes show up. A plage area may eject and suck back a tongue of luminous gas. Or a hydrogen flare may suddenly form in the neighborhood of a large spot group. These flares often cause simultaneous fadeouts of short-wave radio transmission on the earth. For a fuller discussion of these rapid changes, see Chapter 8.

⊙

Fig. 92.   Robert R. McMath and the spectroheliograph of the McMath-Hulbert Observatory.

Narrow-Band Filters   The spectrograph, as a filter, is now being replaced by a new and more effective device. As indicated earlier, we could not use ordinary filters of colored glass because they transmit a band far wider than the narrow spectral line of hydrogen or calcium. There are, however, other ways of separating colors than by spectroscope or

Fig. 93.   The McMath-Hulbert Observatory of the University of Michigan.

stained glass. You must have seen some of them. The rainbow hues of a fragile soap bubble or of a thin layer of oil on a rain puddle result from an optical effect. Certain colors of light are actually destroyed in their passage through the film.

One can produce beautiful color effects in a very simple experiment. Take a piece of cellophane, crumple it, and place it between two sheets or disks of Polaroid. The brilliant color pattern will change kaleidoscopically as you rotate one or both disks. Although the transmitted colors will be as broad as those from a glass filter, this basic principle has been refined.

In place of the cellophane we use a polished optical crystal of quartz or calcite. Other substances will do equally well, as long as they are *birefringent*. Glass, which is monorefringent, will bend a ray of light along a single path. Quartz or calcite, however, will break up a single beam into two rays. Hence the term "birefringent," double bending. These two rays are polarized perpendicularly to one another.

One of these special plates, set between two Polaroid disks, can be adjusted to transmit a series of colors with equally spaced dark

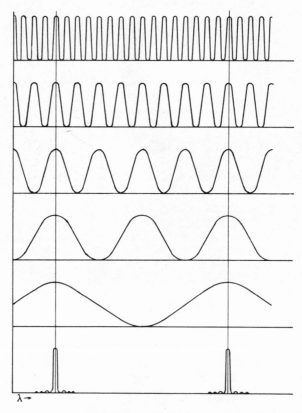

Fig. 94. Transmission curves of layers making up an interference filter. The composite transmission curve of the assembled filter is shown in the bottom line. (Billings, Baird Atomic.)

bands in between. We build up a multidecker sandwich of Polaroid disks and quartz. Each successive layer of the latter is twice as thick as the one preceding. The complete sandwich will transmit a band whose narrowness is determined by the thickest piece of quartz. The thickest layers, for which the size of the quartz crystal may be prohibitively great, are usually made of calcite. Other

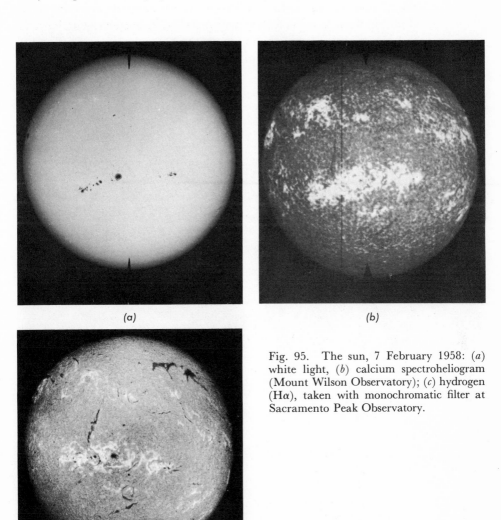

(a)

(b)

(c)

Fig. 95.   The sun, 7 February 1958: (a) white light, (b) calcium spectroheliogram (Mount Wilson Observatory); (c) hydrogen (Hα), taken with monochromatic filter at Sacramento Peak Observatory.

crystals have been used, but none so successfully as quartz and calcite.

Billings and Evans have even designed filters whose passband can be varied, or "tuned," to one of a variety of wavelengths; but the construction and operation of these has proved difficult.

Monochromatic filters enable the astronomer to take direct photographs of the entire solar disk or any portion of it. With the filter we eliminate both the scanning slit of the spectroheliograph, and the moving plate that records the complete image. The full picture usually can be obtained in a fraction of a second.

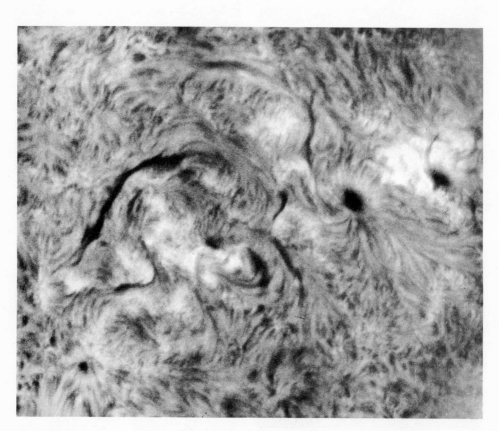

Fig. 96.   Active region near sunspots, in Hα. (Sacramento Peak Observatory.)

With early spectroheliographs the observer could not easily check *Spectro-*
his picture in advance. He could see only a slit image at any one *helioscopes*
time. To overcome this difficulty, Hale devised the so-called
spectrohelioscope. In this instrument an oscillating slit scanned the sun
rapidly enough to produce the effect of a continuous image. One
had merely to make the oscillations fast enough to take advantage
of the persistence of vision—the same phenomenon that makes
motion pictures possible.

The Lyot-Evans filters are simpler to make, however, and, in
general, are less expensive than a spectrohelioscope. They require
no scanning mechanism and can be adapted to any telescope. The
manufacture of such filters does not lie beyond the skill of an
experienced amateur telescope maker.

$\odot$

The monochromatic pictures taken in hydrogen or calcium light *Faculae*
exhibit details that do not appear in unfiltered light. But there are
certain relations between the two types of photographs.

The sunspots, of course, tend to appear dark on all varieties of
pictures, though the calcium photographs often show the spots
completely covered by bright plages. The images in white light also
show bright patches and striations around active spot areas. These
formations, which are most conspicuous near the edge of the solar
disk, are the so-called *faculae* (Fig. 20). Their distribution and
outline correspond so closely to those of the previously described
calcium flocculi that the two phenomena must be basically related.

I have regarded the faculae as a type of solar "mountains." Of
course they cannot be mountains in the terrestrial sense of permanent elevations. But they do seem to represent regions where the
shining photosphere, or solar surface, rises above its average level.
Their close association with spots and their preponderance in the
spot zones lend credence to the upheaval hypothesis. And I have
previously indicated that the calcium plages represent hotter, disturbed, and probably elevated areas of the sun.

Faculae are among the longest lived of all solar features. Although
the markings change with time, faculae often appear before a spot
breaks through the surface, and they remain long after the spot has
disappeared. It seems that we should regard them as being asso-

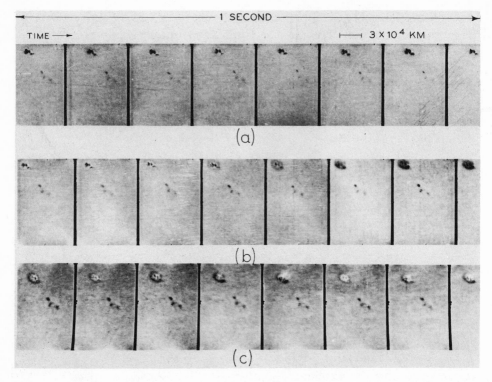

Fig. 97.   Short-lived solar event, 11 August 1954. Exposures (a) show a spot essentially without penumbra; (b), 2 minutes later, shows the rapid formation of a penumbra; (c) 1 minute later, exhibits penumbral brightening and apparent explosive ejection of hot gases. (Courtesy RCA.)

ciated with the most disturbed regions of the solar atmosphere. And, from their great extent compared with the sunspot areas, we conclude that the disruptive forces of the spot zones are deep and widespread. As previously noted, we must attribute them to the enhanced convection expected to occur on the edges of the quiescent spots.

The faculae usually show up in the form of streaks or veins rather than in circular splotches. The lines link together to form series of mountain ranges, rather than isolated peaks, and these veinlike structures tend to be long-lived. The valleys in between are well defined. How high the mountains are is still unknown. They probably extend from 5 to perhaps as much as 100 miles in height. In some cases the bulges may be far greater. The ridges are

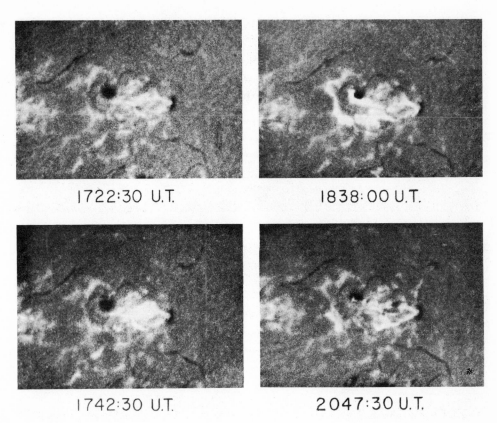

1722:30 U.T.   1838:00 U.T.

1742:30 U.T.   2047:30 U.T.

Fig. 98.   Development of a major flare. Photographs taken in Hα light, between 17ʰ22ᵐ30ˢ and 20ʰ47ᵐ30ˢ Universal Time on 18 September 1957. The small white dot in the lower right-hand corner of the last photograph indicates the size of the earth. (Sacramento Peak Observatory.)

apparently somewhat hotter than the intervening valleys; hence the greater brilliance of the former.

If I were to carry the analogy one possibly unjustifiable step further, I should say that the mountains were violently volcanic. Throughout the entire facular zone we discover large numbers of active prominences. We find geyserlike eruptions, flares, and jets, extending 25,000 to 50,000 miles, and occasionally much higher, above the solar surface. This volume of luminous gas produces the main features of spectroheliograms. Much of the activity cannot of course be described in terms of familiar terrestrial volcanism. We must leave these details, however, until the next chapter.

White-light photographs show that even the areas outside of

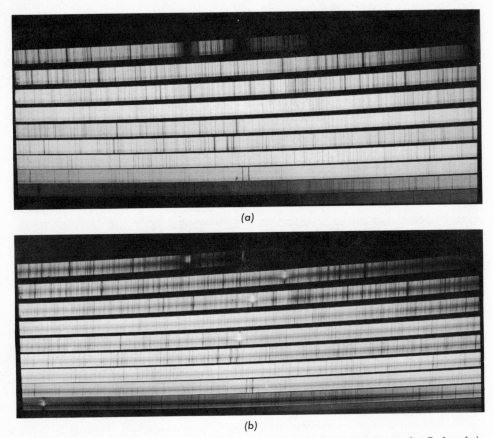

(a)

(b)

Fig. 99.   Solar spectrum, from the violet H and K lines (*top center*) down to the O_2 bands in the far red (*bottom center*): (*a*) normal spectrum; (*b*) flare spectrum. The bright emission lines are caused by the flare of 18 September 1957 (see Fig. 98), the narrow ones are from metals, and the broad fuzzy ones are primarily from hydrogen. (Sacramento Peak Observatory.)

spots and faculae are by no means undisturbed. Minute peaks replace the mountain ranges of the faculae. But the formations change pattern rapidly, within from 5 to 15 minutes. I have referred to the pattern as resembling whitecaps dancing up and down on a stormy sea. It is altogether probable that these minute dots, the solar granulations, underlie the similar, rapidly varying speckles of hydrogen spectroheliograms. The latter may, in turn, be related to the phenomenon of *spicules,* which I shall discuss in the following chapter.

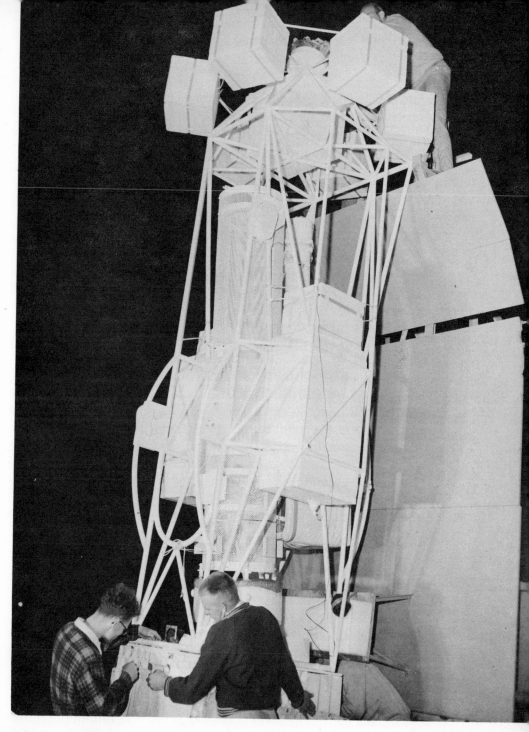

Fig. 100. Telescope and controls for photographing solar granulation from high altitudes, made by Perkin-Elmer Corp. (General Mills, Inc.)

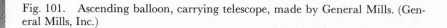

Fig. 101.   Ascending balloon, carrying telescope, made by General Mills. (General Mills, Inc.)

Few observations of granules are available. Those of Janssen, in France, and of Keenan, at Yerkes Observatory, established the average size at 1 to 2 seconds of arc, a range more recently confirmed by Thiessen. From a study of photographs made by Lyot at Pic du Midi, Macris obtains a mean diameter of 1.5 seconds of arc, which corresponds, at the solar distance, to a diameter of about 700 miles.

In Chapter 4 I explained how absorption in the earth's atmosphere complicates the problem of measuring the quantity and quality of the solar radiation. Another effect of the earth's atmosphere interferes with study of features as small as the solar granules. Irregular motions of our atmosphere bend and twist the incoming light rays, usually blurring out the finer details; the same effect causes the stars to twinkle. To obtain better photographs of granules, a group of scientists led by M. Schwarzschild of Princeton recently used giant unmanned balloons to carry a telescope and camera to over 80,000 feet, well above the major part of our atmosphere. From the first three flights, made in 1957, they secured photographs showing the granules with sharper resolution than ever achieved from the ground. The granules show an irregular but cellular appearance—rather like a mud flat, dried and cracked—with diameters from 2 seconds down to 0.3 second of arc, or as small as 150 miles.

While such measures are needed to see the finer details, even a 2-inch telescope will reveal some of the coarser details, which may consist of clumps of granules, rather than individual grains. Under this magnification the solar surface appears mottled, like the peel of a lemon.

H. H. Plaskett has found that the granules give slightly more light in the blue than do neighboring darker areas. This result is consistent with the probable higher temperature of the former. But the observations are difficult and the granules may be even bluer and hotter than Plaskett has suggested. It is extremely difficult to keep the light of neighboring cooler areas from diluting that from the minute bright spots.

Sunspots and faculae vary with the 11-year cycle. They also show abrupt fluctuations from day to day. Do the granules show a

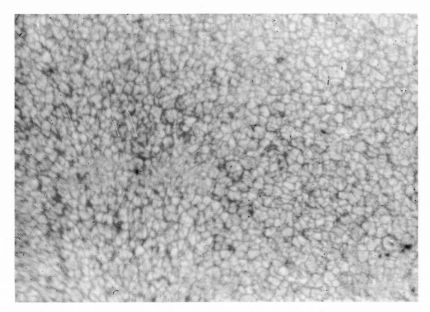

Fig. 102.   Solar granulation photographed by a high-altitude balloon-borne tele-scope. (M. Schwarzschild, Princeton.)

similar variation? A group of amateur astronomers, sponsored by the American Association of Variable Star Observers, are studying this hitherto neglected field. Many observers are needed, if the un-certain effects of "seeing" in the earth's disturbed atmosphere are to be fully eliminated.

Limb Darkening

He who looks at the sun with a small telescope will find clear evidence that the edges are less luminous than the center of the disk. This phenomenon is called *limb darkening*. Early students ascribed the effect to an absorbing action of the sun's atmosphere, but in this conclusion they were only partially correct. The argu-ment was that the thicker layers of atmosphere, near the limb, cut off the sun's rays in the same way that the earth's atmosphere dims a star near the horizon.

We now realize that the phenomenon, although definitely atmos-pheric, is not truly one of extinction. As previously explained, we

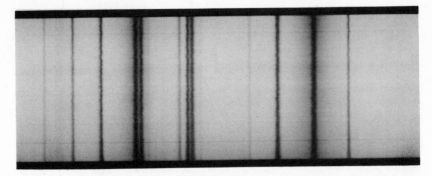

Fig. 103. Wiggly lines caused by Doppler motions of individual granules. The strong line on the right is from Mg at 5172A, and the pair on the left is from Mg and Fe. (McMath-Hulbert Observatory.)

see into the solar atmosphere down to a region where the gas becomes hazy. The temperature of the solar atmosphere—at least for the regions we are now discussing—increases with depth. We see farthest into the sun near the center of the disk and less far near the limb. Thus, the regions near the sun's edge correspond to higher, cooler levels. In consequence they radiate less light and heat than does the center. Violet radiation falls off more rapidly than the red, as the temperature diminishes. Hence, the bluer the light, the darker the limb appears relative to the center.

Early theories, by Lane, Emden, and others, led to the conclusion that the sun's limb should be completely dark—in obvious disagreement with the observations. K. Schwarzschild then suggested that radiation rather than convection transported most of the energy from interior to exterior. Detailed theoretical studies, chiefly by Milne and Lindblad, showed that the darkening and color distribution of sunlight were in good accord with this hypothesis. However, recent studies have shown that convection too does play a significant role in energy transport. As noted earlier, the hydrogen convective zone is effective in the outer one-tenth or so of the solar radius.

Why does the sun's atmosphere become hazy? Or, to rephrase the question: what atoms or substances are responsible for the haze? Suppose, for example, that the sun's atmosphere were similar to that of the earth. Where would the photosphere lie?

The earth's gaseous envelope is really quite transparent. Some scattering of light occurs, to be sure, but we readily see the sun, moon, and stars through it. If the air were about 20 times denser than it is at present, the daylight sky would appear as an almost uniformly bright haze. The sun would still be visible, but perceptibly dimmed. Certainly an observer on the moon would not be able to detect the outlines of continents. The sun's gravitation exceeds that of the earth by a factor of 27 and thus compresses the gas still more. If we could transfer such a superdense atmosphere to the sun, we should find that its pressure there would be 27 × 20 or 540 atmospheres!

From observations of the solar spectrum we easily discover that the actual pressure at the surface of the sun does not exceed one-tenth of our atmosphere. Otherwise, the absorption lines would be much more intense and greatly broadened by pressure effects. We conclude, then, that the solar gases are at least 5000 times more opaque than those of the earth. Why this difference?

The primary reason, as Russell and Stewart showed, is that the solar gases are highly ionized. Electrons attached to atoms are relatively poor absorbers except for specific line radiations. Electrons torn loose from atoms and interacting with the atomic fragments are very opaque. Certainly you have no difficulty in looking across your study. But ionize this gas and you would find that even 10 feet of air possess a high opacity. The electrified gases would be thicker than a London fog.

As for the specific substance responsible, Wildt suggested that hydrogen atoms, of which there is great abundance, capture some of the available free electrons. The resulting negative hydrogen ions, acting conjointly with additional free electrons, prove to be the major source of the sun's atmospheric opacity in the red and infrared. Chandrasekhar carried through the detailed analysis of

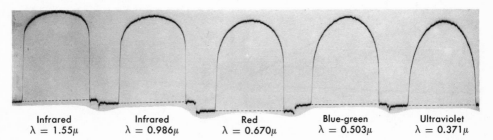

Infrared	Infrared	Red	Blue-green	Ultraviolet
$\lambda = 1.55\mu$	$\lambda = 0.986\mu$	$\lambda = 0.670\mu$	$\lambda = 0.503\mu$	$\lambda = 0.371\mu$

Fig. 104.   Solar limb darkening for different colors. A bolometer, drifting across the solar image, records the distribution of light intensity over the disk. The infrared and red are most nearly uniform. The ultraviolet is most completely darkened. (Abbot, Smithsonian Institution.)

the properties of the negative hydrogen ion. He and Chalonge have independently checked the expected against the observed limb darkening and find good agreement. Recent work by Varsavsky indicates that the hydrogen molecule (H_2) may be a major source of opacity in the "rocket ultraviolet." The opacity of hydrogen in molecules and negative ions thus determines the depth to which we see and hence the location and properties of the photosphere, the luminous outer boundary of the sun.

8

Prominences—Geysers and Volcanoes

The title of this chapter is, in a sense, both inaccurate and misleading. The phrase "geysers and volcanoes" attempts to convey an impression of the general explosiveness of the sun's atmosphere. But the terrestrial prototypes of these phenomena are relatively so puny and weak that no simile could possibly be adequate. Moreover, many of the cloud forms and motions displayed by gases of the solar envelope have no earthly counterpart. "Solar meteorology" may well become an important new science.

The lowest visible layer of the solar globe is the shining surface, the *photosphere*. Above that level lies the cooler gas responsible for most of the absorption lines. We generally call this region the *reversing layer*. The upper fringes of the absorbing region form the *chromosphere*. Lockyer chose this designation because these higher levels, as seen at a total solar eclipse, exhibit a distinct color;

as explained in Chapter 2, we call the bright-line spectrum radiated by these atmospheric gases the *flash spectrum*.

⊙

*Chromosphere
and Spicules*

Observations show that the chromosphere is conspicuously non-uniform. Secchi long ago described it as consisting of tiny fibers that interlace like blades of grass. The word "tiny" applies, of course, only in the solar sense. The individual blades are several hundred miles in diameter and extend upward to heights of 5000 to 10,000 miles. In the equatorial regions they are often inclined or bent as depicted in Fig. 106. Near the poles, however, as W. O. Roberts has shown, they tend to be predominantly radial.

At this point, we must admit to some confusion of nomenclature. Roberts applies the term *spicule* to the individual blades, especially those in the polar regions. Others employ the term to describe the grasslike structure in general. Although agreement has not been reached concerning the detailed behavior of a spicule, I see nothing against using the term in the general sense.

The rates of development and activity of equatorial and polar spicules seem to differ. The latter are considerably more rapid. Roberts, observing the sun at Climax, Colorado, has found the typical polar spicule forming as a sort of blister on the solar surface. The swelling quickly increases in height until it forms a sharp, mountainous peak, ever narrowing at the base. The luminosity of the gas fades rapidly as this jet rises. Simultaneously, the swelling subsides. These spicules complete their cycle in a short time, from about 20 seconds to half an hour, with an average life of from 4 to 5 minutes, according to Roberts. He and Rush have estimated that some 20,000 spicules exist over the entire solar surface at any given moment.

Dunn, observing with the 16-inch coronagraph at Sacramento Peak Observatory, has obtained motion-picture records of spicules under high magnification. He finds that spicules tend to occur in clusters, projecting more or less radially from an elevated hummock, like quills from the back of a fretful porcupine. Some of the larger spicules have a jetlike behavior, spewing matter upward; others, developing into a columnar form with a sharp central core, suggest a searchlight beam shining in a foggy atmosphere. The chromo-

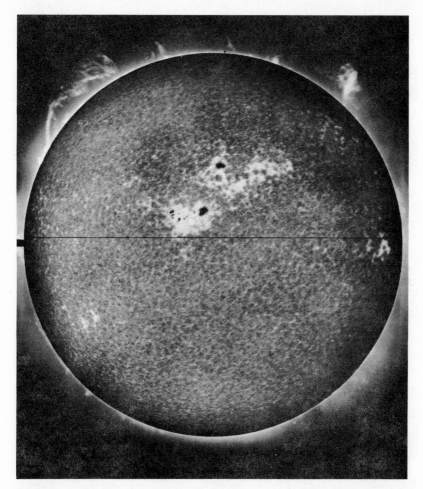

Fig. 105.   Composite calcium (K) spectroheliogram showing the solar surface and surrounding prominences. (Mount Wilson Observatory.)

sphere is in a constant state of flux, with spicules continually forming and fading.

The behavior and structure of the chromosphere suggest a close relation between the spicules and the lower-lying granules (see p. 154). In actual dimensions and frequency on the disk, the two phenomena roughly correspond. If, to extend an earlier analogy, we regard granules as waves on the solar ocean, the spicules are the whitecaps, the spume or spray shot upward by the violent surge of the waves.

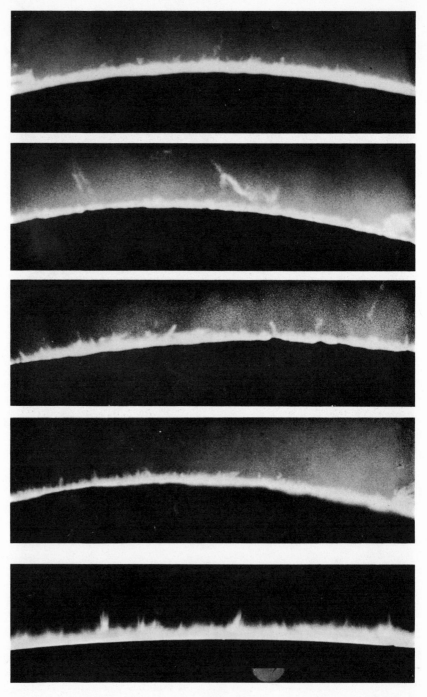

Fig. 106. Chromospheric spicules, photographed at four different eclipses, in 1898, 1900, 1905, and 1918 (Lick Observatory), and (*bottom*) outside of eclipse (Dunn, Sacramento Peak Observatory).

Also, spicules may prove to be fundamentally related to the fine, brushlike formations of the solar corona. In view of the fact that the core of the spicule does not visibly fall back upon the surface, we may at least speculate that the hot spicule material goes to form part of the corona.

The solar chromosphere, then, is not a uniform, quiescent atmosphere. Rather, it is a system of pulsing, interlacing spicules. Most studies of physical conditions in the chromosphere, including my own, have ignored this filamentary structure. The actual densities and pressures inside the jets may therefore be considerably greater than the values previously given for the solar atmosphere. I have derived, for example, an average value of about one ten-millionth of an atmosphere for the electron pressure of the lower chromosphere. The chances are that the figure should be 10 or even 100 times larger within the base of the filaments.

R. N. Thomas has suggested that *shock waves* have much to do with the structure of the chromosphere. Shock waves are intense, highly concentrated pulses of mechanical energy, traveling through the gas at speeds equal to or greater than the speed of sound. A jet plane produces an enormous amount of noise. The plane is accompanied by a shock wave much like that accompanying a projectile (Fig. 107). When this shock wave strikes the earth, it produces a noise like thunder, often shattering windows by its violence. Shock waves may travel at speeds much higher than that of sound.

In the terrestrial atmosphere of nitrogen and oxygen, sound travels approximately 100 feet per second. The velocity is much higher in the chromosphere, of the order of 10 kilometers per second, partly because of the higher temperature and partly because the light element, hydrogen, is the dominant substance on the sun. In the hotter corona, the speed can be even higher, up to 100 kilometers per second or more. Magnetic fields, when present, lend rigidity to the medium and increase the speed of sound still further.

The churning of the vast solar "ocean" sets up a din whose intensity surpasses imagination. Bass notes, or rather subbass notes, predominate. Where 20 cycles per second constitute a buzz barely audible to the human ear, the dominant frequency of the solar atmosphere—corresponding to sound waves 500 kilometers or so in length—is about 1 cycle per minute. Though we cannot hear so low a tone, an observer located on the solar sea would be painfully aware of the vibrations, as waves of pressure or rarefaction follow

Fig. 107.   Photograph of a rapidly moving projectile preceded by a shock wave, through which the temperature and pressure change abruptly. (Avco Research Laboratory.)

one another at intervals of a minute or so, with intensity great enough to break the ear drums.

The blast of noise roaring out through the atmosphere has a pronounced effect upon the outer layers. M. Schwarzschild suggested that such waves might effectively heat the tenuous outer corona, and the high temperature of the sun's outer envelope might result from such action. However, in actuality, the noise is really too violent to be called "sound." The blast is rather a type of shock wave.

Probably only one thing exists more violent than a shock wave, and that is a *focused* shock wave. As Krook and Menzel have pointed out, the wave troughs of the solar seas provide a sort of reflector mechanism, which can converge the shock waves most violently in a region immediately above the point of lowest level. It may well be that the most elevated and excited levels of the chromosphere lie over the troughs in the photosphere.

We should not ignore the possible effects of electromagnetic forces on the structure of the solar atmosphere. Magnetic fields are an observed and well-substantiated phenomenon. The electric currents necessary to sustain these fields are enormous. Moreover, the currents tend to be unstable and can produce effects similar to those observed in prominences. In particular, a current loop will expand in diameter, meanwhile contracting the current-bearing part of the circuit. The magnetic fields associated with these currents must also exert a focusing action upon shock waves traveling through the medium.

⊙

Analysis of
Flash Spectra

Our knowledge of physical conditions in the chromosphere comes largely from studies of flash spectra obtained at total eclipses. Totality begins when the moon just blots out the last traces of continuous spectrum from the atmosphere. At that instant, the higher chromospheric layers show up as a crescent because, for the eclipse to be total, the apparent diameter of the moon must exceed that of the sun. If we place a glass prism in front of our telescope, the resulting spectrum consists of a series of colored crescents, images of the uneclipsed portions of the atmosphere projecting beyond the rim of the moon. Each shade corresponds to a definite chemical element. The crescents are bright, characteristic of the emissions of a gas at low pressure.

Flash spectra show a rough correspondence to the dark-line Fraunhofer absorption spectrum (see p. 85). The same lines appear in both. Close examination, however, will disclose a number of marked differences. Lines of helium and ionized helium, which do not show at all in the normal spectrum, usually appear in the flash. In general, the lines of all ionized metallic elements are far more intense, relative to lines of neutral elements, in the chromospheric spectrum than in the regular dark-line spectrum of the sun.

We could attribute the increased ionization either to lower pressure or to higher temperature. Astronomers, believing that the rim of the sun should be cooler than the lower levels, were long inclined to accept the former explanation. At the lessened density of the upper atmosphere, as Saha, the Indian physicist, first pointed out, the atoms would remain ionized longer, because electrons for capture are scarcer. This theory accounts satisfactorily for the character of most of the flash spectrum.

The occurrence of helium and ionized helium, however, cannot be so explained. I have previously mentioned that helium is particularly difficult to excite. The atoms giving the helium lines will not shine or absorb unless the effective excitation corresponds to 20,000° K or more.

There are two ways for exciting a helium atom—or any atom, for that matter—collision and radiation. A sufficiently energetic particle smashing into an atom can make it radiate energy. Light of appropriate wavelength will produce the same result.

We can readily see why lines of calcium, iron, or sodium should be intense in the chromospheric spectrum. At temperatures of from 4000° to 6000°, many electrons and light quanta are available to raise these atoms to an excited level. Helium and ionized helium, however, require from 10 to 20 times as much energy as the ordinary metallic atoms. One easily calculates that the temperature needed to produce enough excitations, by either collision or radiation, is 20,000° K at the very least.

To complicate the problem still further, we find that helium excitation increases toward the upper atmospheric levels. Here is a seeming impasse. All of our preconceptions would lead us to believe that the outer solar layers should be cooler than the inner ones. The observations, however, force us to the opposite conclusion, that the upper levels are hotter. Perhaps, as Athay and Menzel have suggested, we have in the chromosphere alternate columns of cold and hot gases, the latter heated by shock waves, as described in the previous section.

⊙

Solar
Prominences   Solar prominences, which are flamelike emanations of the chromosphere, come in a variety of sizes and types. Some appear to be little more than enlarged spicules. Others display peculiarities that require detailed description.

On 4 June 1946 there occurred one of the greatest solar explosions in history. The photographs in Fig. 108 exhibit strikingly what happened.

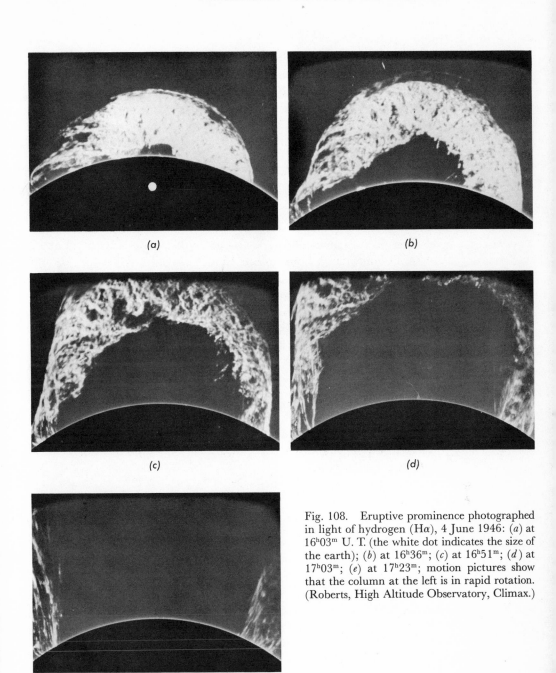

Fig. 108. Eruptive prominence photographed in light of hydrogen (Hα), 4 June 1946: (a) at 16ʰ03ᵐ U. T. (the white dot indicates the size of the earth); (b) at 16ʰ36ᵐ; (c) at 16ʰ51ᵐ; (d) at 17ʰ03ᵐ; (e) at 17ʰ23ᵐ; motion pictures show that the column at the left is in rapid rotation. (Roberts, High Altitude Observatory, Climax.)

The first, taken shortly after sunrise, shows a great arch of fiery gases rising above the curved edge of the sun. The sun's bright surface is hidden behind the black disk. For comparison, the size of the earth is indicated as a white dot at the bottom of the picture.

The eruption was already under way when the first picture was obtained. Thirty-three minutes later the prominence had risen to a height of 250,000 miles, the outer boundary of the arch being almost a perfect arc of a circle. The second picture clearly shows that the material of the prominence consisted of myriads of filaments and streamers, with occasional knots and condensations. These streamers seem to be wound in a coil, resembling a huge spring.

The prominence continued to rise rapidly with a speed of about 400,000 miles per hour, more than 100 miles a second. In the fourth picture, the top of the arch has moved entirely out of the frame; in the last photograph, taken only 1 hour 20 minutes after the first, the material has blown far above the range of the instrument. The ring eventually expanded outward to well over a solar diameter. It is the size rather than the behavior of this prominence that is so striking. Small ascending arches are not uncommon, but the tremendous force of this solar explosion is unprecedented.

These vast clouds of gas, which appear luminous beyond the solar boundary, are the formations that show as the bright and dark flocculi on the disk. The very fact that some of the clouds show dark in absorption and others bright in emission emphasizes their variety.

⊙

Prominence

Classification   Early observers divided prominences into two classes: quiescent and eruptive. Modern records, especially those taken with motion-picture equipment, indicate that this classification is inadequate. Pettit and McMath have extended the system. Pettit's latest revision follows:

Class I	Active	Ia	Interactive
		Ib	Common active
		Ic	Coronal active
Class II	Eruptive	IIa	Quasi-eruptive
		IIb1	Common eruptive
		IIb2	Eruptive arch

Class III	Sunspot	IIIo	Cap
		IIIa	Common coronal sunspot
		IIIb	Looped coronal sunspot
		IIIc	Active sunspot
		IIId1	Common surge
		IIId2	Expanding surge
		IIIe	Ejection
		IIIf	Secondary
		IIIg	Coronal cloud
Class IV	Tornado	IVa	Columnar tornado
		IVb	Skeleton tornado
Class V	Quiescent		
Class VI	Coronal		

Some of these categories require explanation. Interactive prominences consist of two or more basic prominence spikes, which appear to exchange material, knots, or streamers. The common active prominences, according to Pettit, are the most abundant of all types. They consist of a single mass of tangled filaments, with knots in rapid motion. The direction of gaseous flow is generally uniform, toward some area that Pettit terms a "center of attraction." The speeds of motion are surprisingly constant.

Occasionally a streamer of luminous material moves in from the outer coronal regions. Pettit terms such formations "coronal prominences" or simply "coronals." When one of these coronals enters an "attraction center," near a Ib prominence, we call it Class Ic. Otherwise it is of Class VI.

Downward motions prevail in all Class II prominences. The knots and filaments brighten markedly as they approach the photosphere, where perhaps the gases encounter some slight resistance or are excited by electromagnetic effects. Occasional fragments detach themselves from the tips of the prominence and rise, slowly fading in intensity. Some of these resemble comets, with a fanlike tail pointed away from the sun.

Active prominences frequently develop into the eruptive Class II. The continued activity often causes the prominence to expand in the form of an arch. I suspect that the primary differences between the subclasses in Class II are partly of degree and partly of orientation. A filamentous arch viewed from the end looks like a towering tree. Or, if one end of the arch breaks loose and flies upward, we have a formation similar to IIb.

Pettit states that his Class III prominences occur only over spots, although spots frequently have no prominence appendages at all. Roberts, at Climax, reports that he has frequently found prominences whose form and activity are identical with those described as Class III, but where there is no trace of an underlying spot. Nevertheless, these characteristic prominences seem confined pretty well to the sunspot zone. Some of them may be closely associated with the bright calcium flocculi, which are similarly distributed. The flocculi, for example, are brightest around the disturbed spot areas. But condensations do occur elsewhere. Perhaps their spots are buried, with the prominences and flocculi the chief visible indications of a subsurface activity.

Under Class III, Pettit lumps a number of prominence varieties whose characteristics show a wide difference. The cap prominence is merely an elevated region of the photosphere. That the rise comes about from some internal pressure is indicated by the existence of occasional surges from the disturbed area. Surges, Class III, are ribbons of material ejected at very high velocity. The material reaches a maximum height and then, with fading intensity, falls back into the sun. One has the impression that the sun often "sucks back" the flamelike tongue along the original path.

To the filamentary surge, Pettit adds Class IIId2, the expanding surge, which breaks into a sort of spray. There are other possible classes, not included by Pettit. One is a surge with a mushroom cap, resembling an atomic explosion. Still another and possibly extremely important type, not yet included in the classification, is the semi-invisible surge. These last are evident from the effects they produce. One sees merely the beginning—a bulge and a burst. But I believe that they may eject a vast amount of invisible material.

The coronal spot prominence, Class IIIa, consists of a series of streamers moving downward, forming a fanlike pattern. Pettit terms IIIb the "looped coronal sunspot prominences." I should prefer to drop the adjective "coronal," because the activity is not necessarily related to the corona. These typical looped prominences usually start as a bright arc well above the surface. The area of luminosity grows downward in two directions, forming a well-defined loop. The material, surprisingly, appears to flow downward in both sides of the loop. There is no obvious source of this material. The mystery is as great as if one held both ends of a loop of disconnected hose in his hands, and found water pouring out in both directions!

Except for scale, prominences superficially resemble terrestrial

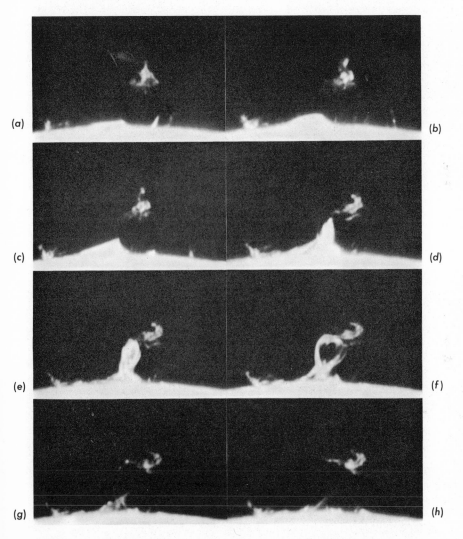

Fig. 109.   Remarkable development of a surge into the form of a loop, Hα spectro-heliograms, 23 September 1938: (a) 18^{h}03^m U.T.; (b) 19^{h}13^m; (c) 19^{h}43^m; (d) 19^{h}48^m; (e) 19^{h}54^m; (f) 20^{h}01^m; (g) 20^{h}08^m; (h) 20^{h}11^m. The disconnected patch of luminosity is probably independent of the explosion. (McMath-Hulbert Observatory.)

Fig. 110. A geyserlike surge rising vertically to about 250,000 miles. Thereafter the eruption faded, but most of the material was falling back into the sun. (High Altitude Observatory, Climax.)

Fig. 111. Giant loop prominence. (Sacramento Peak Observatory.)

174

clouds. Both are wispy or fibrous in structure. The visible material is constantly changing, condensing on one side and evaporating on the other.

Pettit's classification is primarily descriptive of the form or character of a prominence at any moment. Class V may turn to Ia or Ib and then go into an eruptive Class II. One of the fragments of the eruption may be classified as IV, and so on. Thus, there is nothing generic about the classification, valuable as it is for taxonomic purposes. The varieties listed under III, however, seem to be distinct.

Menzel and Evans, from a detailed analysis of motion-picture records from Climax and Sacramento Peak, have devised a so-called "behavior" classification system of solar prominences. They recognize two basic classes, one where the luminous material comes primarily from above (A), and another where it originates from below (B). Each class further subdivides into two groups, prominences associated with spots (S) and prominences not so associated (N).

The complete classification, with some modifications suggested by Orrall, is as follows:

A. Prominences originating from above in coronal space
   S. Spot prominences
      a. Rain
      f. Funnels
      l. Loops
   N. Nonspot prominences
      a. Coronal rain
      b. Tree trunk
      c. Tree
      d. Hedgerow
      f. Suspended cloud
      m. Mound

B. Prominences originating from below in the chromosphere
   S. Spot prominences
      s. Surges
      p. Puffs
   N. Nonspot prominences
      s. Spicules

The names attached to the different classes are intended to convey a rough picture of the general appearance of the prominence. Those

Fig. 112.   Large-scale picture of a giant loop. (Sacramento Peak Observatory.)

of type *A,* where the material seems to condense from above and fall back to the solar surface, are far more common than those of type *B.*

⊙

*Physical
Nature of
Prominences*   Motion-picture records tell us far more about the nature of prominences than do isolated pictures.

Let us ask this question, which seems to be fundamental. Can we distinguish between the prominences that show up as bright and as dark flocculi on the disk? Is there some fundamental physical difference between the two sets, so defined? We have generally identified absorption with the larger prominences and emission with the smaller ones—as if total quantity of material were responsible. The problem may, however, be more involved.

Fig. 113. Hedgerow prominences, photographed with a large coronagraph. (Sacramento Peak Observatory.)

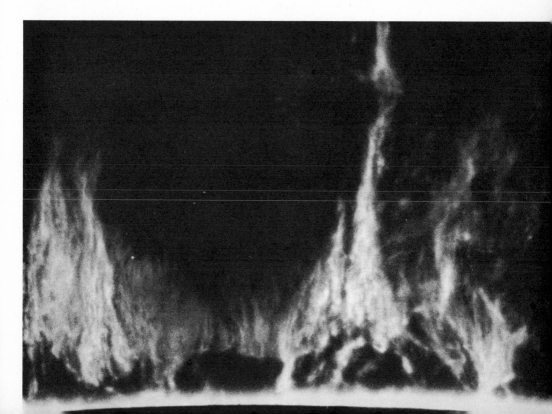

We have already noted that an atom emits radiation as it changes from one pulsating pattern to another (p. 56). In hydrogen, we may label the patterns by consecutive numbers, 1, 2, 3, and so on, beginning with the one of lowest energy. The line Hα results from transitions between patterns 2 and 3 of the hydrogen atom. The jump 2 to 3 represents absorption; the reverse produces emission of radiation. If the two transitions exactly balanced, the prominence would be neither bright nor dark. It would show, if at all, as only a minor blemish on the solar disk.

The bright prominences require a greater proportion of hydrogen atoms in level 3, a condition that implies a higher temperature. Thus, I conclude that the bright flocculi are hotter than the dark flocculi. From the standpoint of possible terrestrial effects of prominences, the foregoing distinction is especially important. A bright region should emit far more ultraviolet radiation than does a dark area. The ultraviolet is responsible for the ionization in the upper levels of the earth's atmosphere, so that a classification based on the distinction between bright and dark patches may be useful from the standpoint of terrestrial effects. However, only the larger and more spectacular prominences show up distinctly in projection against the disk, making such a classification difficult.

At this point I would note that a still older division of prominences into two classes—hydrogen and metallic—may possibly be related to the respective classes, dark and bright. But there is almost certainly no basic chemical difference between the two groups. The distinction is largely one of intensity. A prominence, however large it may be, with a low surface brightness will show readily only the hydrogen lines, the yellow line of helium, and the violet lines of ionized calcium. The fainter metallic lines of magnesium and ionized iron are there, but drowned out in the glare of the sky.

A prominence with high surface brightness, however, will show additional lines. It may even happen that the hydrogen and calcium lines will have reached the intensity of saturation, so that the metallic lines mentioned above are relatively brighter. Thus no real distinction exists, except surface brightness and possibly density, between a metallic and a hydrogen prominence. But the prominences with intense emission are those most likely to show bright on the disk.

The dark hydrogen flocculi are among the most outstanding

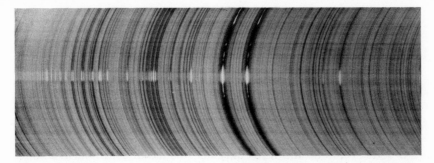

Fig. 114.   Spectrum of a bright solar prominence, 6 September 1957, from the head of the Balmer series around 3640 A (*left*) to beyond Hδ, 4102 A. The brightest lines are H and K of calcium. (Orrall, Sacramento Peak Observatory.)

features of disk spectroheliograms, projected against the bright solar surface (Figs. 84–86). These objects are, for the most part, of class ANd. Their edges are irregular, scalloped or serrated. One frequently hears these objects referred to as "dark filaments." However, we must not visualize them as uniform clouds. I have already mentioned the fact that most prominences consist of delicate, interlacing threads. The mass resembles a strand of wool yarn, in which the fibers occupy only a small fraction of the total volume.

The comparison to yarn is apt in more ways than one. The diameter and shape of the prominence threads vary markedly from place to place. Here the strand is regular; there it is flattened and torn. Fibers connect it with the solar surface. Probably the most common form is a sort of flat, filamentous ribbon, its length parallel and its width nearly perpendicular to the solar surface. Material seems to be flowing from above into the region, the filaments condensing more and more, the nearer they approach to the solar surface.

When one of these dark filaments passes over and around the solar limb, we see it as a bright prominence. When it appeared against the disk, we could not visualize its vertical dimension. Now, we see it edge-on, where it curves around the disk, and we cannot easily picture its extension toward or away from us.

Records of such a prominence often display a striking similarity of form for several days, despite the fact that the sun may have turned through an angle of 45° to 60°. The cross section of the filamentous ribbon may tend, in some prominences, to be similar at various places along its length.

The motions of the gases in objects of this type are remarkable. We can trace such activity by observing the paths of individual knots and condensations in the filaments. One of the surprising features of these motions, first pointed out by Pettit, is the constancy of the velocities. Here and there, to be sure, we find sudden accelerations. But there is no sharp increase of speed such as would be expected for matter falling in the gravitational field of the sun. In fact, if gravitation were the only force acting, matter would cascade from a height of 20,000 miles to the solar surface in an interval of only 8 minutes, acquiring a velocity of 80 miles per second during the descent.

It is a matter of interest to determine how clouds can remain suspended, day after day, without seeming change of form. In many cases, however, we see part of the answer. The form does not alter, but the matter within the prominence is not the same. The effect is similar to that of a fountain, whose figure remains constant, though the water flows continuously.

The great preponderance of downward motions is one of the most striking features of prominences. Where does the matter come from? Why do we not see it moving up?

We cannot give a complete answer to such questions at the present time. But the nature of the probable answer may appear from a familiar analogy. On the earth, rain falls down, never up. I can imagine that the more inquisitive of our prehistoric ancestors may have puzzled over the problem. They postulated the existence of rain gods, whose duty it was to fill the skies with water. Unfamiliar with the phenomenon of evaporation, the ancients failed to realize that the water went skyward in a different physical form.

The solar "rain" consists of luminous atoms. Perhaps the material moves upward in a nonluminous form. A gas may fail to shine if its temperature is either too low or too high. I cannot definitely exclude the former possibility at the present time, though the evidence is somewhat against it. Certain atoms, like sodium, for example, will continue to shine at very low temperatures. Motion pictures taken in the light of sodium, unfortunately, do not exist. When such pictures are taken, we shall have a means of detecting clouds of cooler gas. However, I scarcely expect to find evidence for cool clouds. The eruptive forces necessary to eject the gases to the great heights required seem characteristic of high rather than of low temperatures.

But we require extremely high temperatures in order to suppress the radiations of hydrogen or helium. The values range upward from a minimum of 500,000°C. The fact that such temperatures appear to exist in the solar corona gives some support to this conclusion.

The dark flocculi, then, appear to represent regions where the gases pass from a state of high to one of lower temperature. The phenomenon results from a general circulation involving the entire solar atmosphere, including the corona. The floccular formation may be somewhat analogous to what the meteorologist terms "fronts" in the earth's atmosphere. On earth, we encounter a variety of types of "air masses," such as the cold, dry polar masses, and the warm, moist tropical masses. A front is the dividing line between two independent air masses.

The characteristics of sunspots indicate the presence of regions of low pressure. We expect to find some compensating regions of high pressure. We should not stretch the terrestrial analogy unduly. Nevertheless, if there existed two contiguous areas of high pressure, with a low-pressure valley between them, one might expect a circulation of the type observed. The filamentous condensation would form in the valley. The fact that the filaments often point toward the spots, and occasionally appear to be sucked into the spots, is additional evidence in favor of the foregoing interpretation of their physical origin.

Giovanelli has suggested that filaments may form in regions where the magnetic field of a spot just neutralizes the general magnetic field of the sun. The electrical conductivity of a gas is strongly dependent upon the presence or absence of a magnetic field. A strong field tends to lower the conductivity. Thus electrons might find a race track along the paths where magnetic forces just cancel one another. The picture is far from complete, but it does contain a suggestion that may be profitable for further studies. Giovanelli applies his theory especially to the production of solar flares. The ribbon formations, in my opinion, are even more likely to arise from such a magnetic situation.

The filamentous ribbons account for all but Class III in Pettit's classification. Many if not most of the quiescent prominences fall into the filamentary category. The active and eruptive varieties represent evolutionary stages in the same basic variety. The so-called coronal prominences may be the threads of an incipient ribbon. The sunspot prominences may possibly be—though I am somewhat doubtful—loops in the ribbon or in the threads of the ribbon.

Motion pictures show that the material in a filament is constantly being replaced. Threads form in the upper regions and weave their way through to the solar surface. Successive threads tend to follow similar paths, as if there were preferential regions for the condensations to form.

We know very little about the early stages of filament formation. From very meager evidence, I conclude that the newly formed prominence is relatively inactive. The motions of the gases are slow. Later in life, the activity increases. Knots and condensations occur more frequently and move with greater speed. Then, in its final stages, the prominence rises far above the solar surface, forming what appears to be a giant arch.

I have already described Fig. 108, which depicts the successive stages in one of these "eruptive" arches—the largest on record. Roberts obtained these records on 4 June 1946, with the coronagraph at Climax, Colorado. Figure 115 shows three views of the same object viewed as a dark hydrogen flocculus against the disk. D'Azambuja, of Meudon, kindly provided these earlier photographs. In the disk pictures, the filament structure is not clearly evident. The prominence lies in so high a solar latitude (southern hemisphere) that we see a considerable portion of its flat side. During the process of elevation, we gain the impression of a sort of spiral structure, as if the prominence had a tubular form. Either that, or the object is curved lengthwise, like a cigarette paper in the process of hand-rolling. The records of most prominences would favor the latter interpretation.

We have generally called such prominences "eruptive," when we discover them so expanding. Much speculation has occurred regarding the nature of the forces that drive the material upward. The most common view has been that *radiation pressure*—the force of sunlight itself—is responsible. Milne and others have built extensive theories of prominences based on such assumptions. Qualitatively, the suggestion appeared promising. We have the supporting evidence of the apparent action of radiation pressure upon tails of comets, whose filmy trains always point away from the sun.

When we subject the radiation-pressure hypothesis to rigorous examination, however, our doubts begin to grow. These prominences

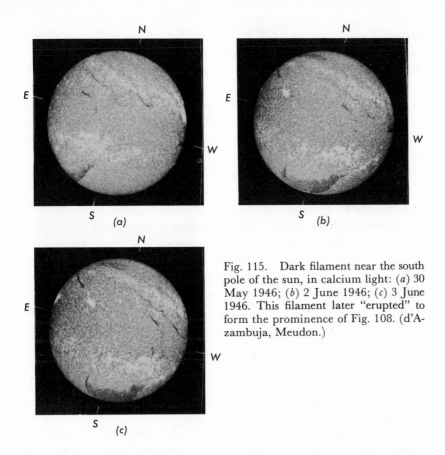

Fig. 115.   Dark filament near the south pole of the sun, in calcium light: (*a*) 30 May 1946; (*b*) 2 June 1946; (*c*) 3 June 1946. This filament later "erupted" to form the prominence of Fig. 108. (d'Azambuja, Meudon.)

possess extraordinary masses. A single spicule may weigh a million tons. A large filament may contain more than 1000 million tons of material. The repelling power of sunlight is simply not great enough to lift or eject this amount of gas—at least without assistance from other forces.

Pettit has pointed out the fact that a balloon moves with constant velocity through the earth's atmosphere, expanding gradually as it ascends. The uniform rate of rise of the solar prominences suggests a somewhat similar action. I am likewise inclined toward a hydrodynamic or possibly an electromagnetic explanation for the motions of solar prominences.

The motion pictures of prominences often suggest that the variations are not of matter but of regions of excitation. An analogy will show what I mean. A searchlight, playing on the clouds at night, will produce a luminous spot. As the beam moves, the position of the luminosity changes. Or, again, the aurora borealis shows rapid variations of pattern that we must ascribe to fluctuations of excitation rather than to motions of the gases in the upper atmosphere.

We find at least a partial answer to this question in observations of the Doppler effect in prominence spectra (p. 50). Remember that the radiation comes from atoms of the gas cloud. If the cloud is stationary and if the changes of pattern result from variations of excitation, we should find the gas quiescent. There should be no marked displacements of spectral lines to the red or violet. But actual observations often show shifts of the lines, indicating that in fact the matter is in violent motion.

Of primary importance for such studies is the type of observation made at the McMath-Hulbert Observatory with their Stone spectrograph (the gift of Julius Stone). This instrument, a special variety of spectroheliograph, takes a series of records across the prominence. Each separate picture notes the position of the basic spectral line. A shift of the line to red or violet means that the photographed portion of the prominence is moving toward or away from us. Each picture thus measures the speeds of gases in the direction of the line of sight. Since the successive frames portray the motions from side to side, we can build up a three-dimensional picture of the prominence motions and ultimately determine the true distribution of material within the body of gas.

Although such observations indicate that the matter is often in violent motion, we cannot immediately conclude that the variation of excitation is entirely absent. On the contrary, we have already attempted to explain prominence regions in terms of temperatures and pressures conducive to the emission of hydrogen and helium radiation. If these conditions change level in the atmosphere, we should expect prominences to rise or fall accordingly. Some of the effect of the expanding prominences may arise from a cause of this sort. Perhaps the extreme activity of the filaments, just prior to the eruption, has filled the low-pressure trough and the prominence rises as a direct effect of gas and magnetic pressure from below.

The peculiar behavior of a loop prominence, which sends down two curving filaments from a luminous seed high above the surface, may well be a spread of excitation rather than of moving matter. But further records are needed to determine the exact nature of this phenomenon.

In this discussion of prominences, we have become increasingly aware of difficulties. There are problems of interpretation that puzzle us greatly. Among them is the prevalence of downward motion. Where does the material come from? What causes it to shine? Why does it move in such peculiar fashion? What supports the clouds of gas against the force of gravitation?

Our studies of the solar surface led us to consider the prominences. Our examination of prominences has carried us out into the regions of the corona. We shall find it profitable to examine these outer regions with reference to the sun as a whole. Then and only then can we conjecture some possible answers to the queries raised in the previous paragraph.

9

The Corona Mystery

The sun's corona is the pearly halo that becomes visible at the time of a total solar eclipse. The ethereal beauty of the corona arises, in part, from the delicate structure of its fanlike rays. Its color is similar to that of sunlight—predominantly white.

The shape of the corona is by no means constant. At times of sunspot minimum, the rays show enormous extensions along the equator, with short brushlike tufts near the poles. In 1878, Langley observed one faint streamer out to 12 solar diameters. At sunspot maximum, the equatorial extension is much less pronounced, and the corona is more regular, its outline like a dahlia.

Coronal photographs from fourteen different eclipses appear in this volume. The mean sunspot numbers, for the month in which each of the eclipses occurred, are listed in Table 4. The corona at the eclipse of 1937, with the highest associated spot number, is

Table 4. SUNSPOT NUMBERS AT TIMES OF ECLIPSES, AND
FIGURE NUMBER OF CORRESPONDING CORONAL PHOTOGRAPH.

Month	Year	Spot number	Figure number
January	1898	30	116
May	1900	15	117
August	1905	59	118
January	1908	39	141
June	1918	59	142
September	1922	5	143
September	1923	13	144
January	1926	72	145
October	1930	34	146
August	1932	7	147
June	1936	70	24
June	1937	130	148
February	1952	23	149
June	1954	0	150

clearly the most jagged and nonuniform, although that of 1905 is almost as irregular. The minimum coronas of 1900, 1922, 1923, 1932, and 1954 show the greatest equatorial extensions and the best-developed polar brushes.

The most spectacular differences between maximum and minimum coronas, however, appear in the drawings by visual observers. Unless photographs are subjected to special printing processes, the appearance of the picture does not agree with the impression recorded by the eye. The eye is sensitive over a very wide range of light intensity; the photographic plate has a much narrower response. Thus, in pictures we ordinarily miss the delicate structure of the bright inner corona, because of overexposure. The special printing techniques originated by Gardner (see Fig. 148) are particularly effective in overcoming this difficulty.

The corona, like the prominences, appears to possess filamentary structure. The inner corona shows domes and arches as well, studied particularly by J. A. Miller and more recently by E. Bugoslavskaya. The 1937 corona (Fig. 148) possesses streamers that seem to diverge, recross, and diverge again. This apparent focusing action strongly suggests the presence of forces like those that a series of overlapping magnetic fields would exert on a beam of electrons. The greatest extensions of the corona, moreover, generally occur for those latitudes where the sunspots are most numerous at the time of observation.

Fig. 116.   Corona, 22 January 1898. (Lick Observatory.)

While the eye in some ways gives a truer impression of the over-all corona, it is in other ways a poor judge of coronal brightness because it tends to exaggerate small differences. If, even in the most jagged-appearing corona, we determine an exact contour of equal intensity (a so-called *isophote*) the line is surprisingly round and symmetrical. The most brilliant streamer shows only as a slight bump in the otherwise smooth contour. Thus the corona is actually surprisingly close to globular in form.

When our study of the corona was limited to total eclipses, with a maximum average duration of 2.9 minutes per year, knowledge accumulated slowly. We could not see the day-to-day variations and we could only guess that the corona was rotating.

The invention of the coronagraph changed this picture. Daily observation became possible and we could investigate problems that hitherto had seemed impossible of solution. The coronagraph is the invention of the distinguished French astronomer, B. Lyot. In developing this device, Lyot brilliantly and patiently overcame,

Fig. 117. Corona, 28 May 1900. (Lick Observatory.)

one by one, all of the difficulties that had led many astronomers to state that the corona could not be studied except during a total solar eclipse.

The problem seems simple enough, at first sight. The corona is about as bright as the full moon and, since the areas of the two objects are roughly the same, the surface brightnesses are comparable. In fact, the inner corona possesses an appreciably brighter surface than the moon. We have no difficulty in seeing the moon in the daytime. Why, then, can we not see the corona?

Part of the difficulty comes from the fact that another sort of brilliant halo, produced in our own atmosphere, ordinarily surrounds the solar disk. Dust and other impurities in the air scatter sunlight to produce this glare. To see it, try a simple experiment. Close one eye and hold your thumb up at arm's length, to obscure the sun. Examine the sky in the solar neighborhood. Unless sky conditions are exceptional the white halo will dazzle you with its brilliance. Only at high altitudes, where rain frequently washes the air, will the sky be blue and dark up to the sun's edge. Such a sky is the first requisite for successful operation of a coronagraph.

The lenses of the instrument impose the second condition. They must be free from scratches, bubbles, or other imperfections. And they must be kept scrupulously clean. Have you ever driven a car against the setting sun? The glare comes not so much from the sun directly as from the sunlight scattered by the dusty or scratched

Fig. 118.   Corona, 30 August 1905. (Lick Observatory.)

windshield. Or have you ever seen a photograph where the camera had been pointed too near the sun? The streaks and spots come from light scattered in the camera and reflected by the faces of the lenses.

Lyot found it necessary to use a simple objective lens because the compound achromatic lenses commonly used for large telescopes cause reflections between the faces. The focal length of a simple lens for blue light, however, is appreciably less than that for red light. Narrow-band filters before the camera, which permit only a small range of colors to reach the eye or the photographic film, are used to eliminate the difficulties of focus that a single lens would ordinarily cause.

The principle of the coronagraph is relatively simple. (Fig. 119). The first (objective) lens forms an image of the sun and corona. A disk at the focus obscures, or artificially eclipses, the image of the solar disk. Then a second (camera) lens focuses the disk and the unocculted light of corona and prominences upon the film. Auxiliary lenses and diaphragms remove the remaining scattered sunlight, much of which comes from the edges of the objective lens.

The filters used with the coronagraph are, in themselves, a very important scientific development. R. W. Wood, Y. Ohman (of

Sweden), B. Lyot, and J. W. Evans are primarily responsible for these superior gadgets. In its simplest form, the filter consists of alternate layers of quartz and Polaroid, as described in Chapter 7. More complex filters can be made with transmission bands so narrow that the instrument, like a spectroheliograph, will give monochromatic pictures of disk or prominences.

The coronagraph, used with filters or a spectrograph, has opened up new fields of solar research. For example, the motion-picture

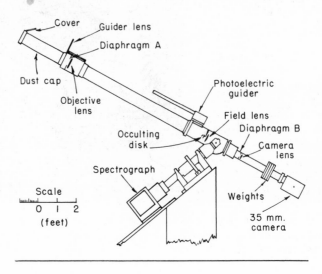

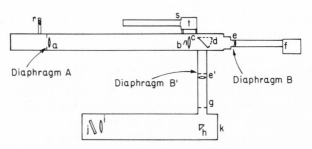

Fig. 119.   Principle of the coronagraph. (*Above*) Form of the instrument. (*Below*) Arrangement of optical parts (schematic): (*a*) objective lens; (*b*) occulting disk; (*c*) field lens; (*d*) swinging mirror; (*e, e'*) camera lenses; (*f*) motion-picture camera; (*g*) spectrograph slit; (*h*) reflecting prism; (*i*) collimator-camera lens of spectrograph; (*j*) diffraction grating; (*k*) focal plane of spectrum; (*r*) guider objective; (*s*) occulting disk for guider; (*t*) photocells of guider unit; (*A, B, B'*) diaphragms to remove scattered light. (High Altitude Observatory, Climax.)

Fig. 120. Coronagraph building at Sacramento Peak Observatory. This large turret dome houses the large 25-foot spar, which carries the most powerful of the optical instruments at the Observatory. Spectrograph wing, lower left.

records obtained by Lyot at the Pic du Midi, by Waldmeier in the Swiss Alps, and by observers at Climax, Colorado and Sacramento Peak, New Mexico, have added a wealth of information concerning the activity of prominences (Chapter 8).

Skellet was one of the first to apply television principles to scan the solar image and throw away the superposed sky glare by electrical means. The image is again rebuilt electrically on a television image tube, and finally photographed. A modern television camera now replaces the mechanical scanner of the original instrument. The complete device, called a "Lumicon," has been used successfully by Evans and his associates at Sacramento Peak to detect the faint corona or coronal spectrum in the presence of strong sky glare.

Let us turn now to consider some coronal problems. The spectrum of the corona consists of a continuum and superposed bright lines. As Moore and Grotrian have shown, there seem to be two continua.

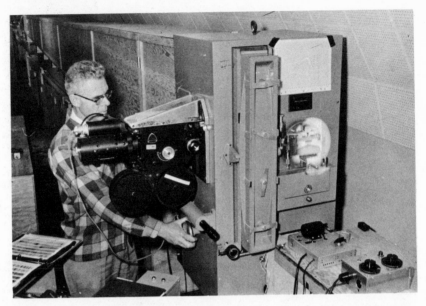

Fig. 121. John W. Evans, Director of Sacramento Peak Observatory, and the large coudé spectrograph.

The outer continuum appears to consist of scattered sunlight. One at least can recognize the Fraunhofer lines, though they show up only as highly broadened and washed-out features. The lower corona shows a pure continuum, in which none of the familiar solar lines appears. Even the intense H and K lines are absent.

Photographs of the corona show that part of the radiation is polarized. The simplest method for recording this phenomenon involves the placing of a disk of Polaroid or other polarizing material just in front of the lens or, better, just ahead of the film. In 1937, a Peruvian amateur astronomer, Fernando de Romaña, obtained for me the photographs that appear in Fig. 124. The arrows show the direction of the Polaroid axis. Note how the form of the corona changes with the direction of the axis.

$\odot$

The Coronal Lines

The emission lines seen in the spectrum of the corona were for many years one of the greatest mysteries in science. The wave-

Fig. 122. Patrol instrument at Sacramento Peak Observatory. This mounting carries a 6-inch coronagraph for motion pictures of prominences and the corona through a birefringent filter, a flare patrol which photographs the whole solar disk in Hα light, and a white-light sunspot telescope.

lengths of these radiations simply did not coincide with any of those from known chemical elements. For a considerable time astronomers postulated the existence of a new chemical element— coronium. The identification of helium had suggested that such a solution was possible. Many years earlier, astronomers had found, in the spectrum of prominences and the chromosphere, a mysterious yellow line of unknown origin. They called the supposed element that produced it helium, after the Greek *hēlios* for sun. Finally, Ramsay found a hitherto unknown gas which gave out identical radiation, and thus showed that helium existed on earth as well as on the sun. But, as the years passed, the chemists filled in all the likely gaps of the periodic table of Mendeleev (p. 52). Thus scientists were forced to conclude that "coronium" was some well-known and probably abundant substance, masquerading under altered wavelengths owing to the peculiar physical conditions obtaining in the solar corona.

In 1940, B. Edlén, a Swedish physicist, solved the problem com-

Fig. 123.   High Altitude Observatory of the University of Colorado, Climax, Colorado.

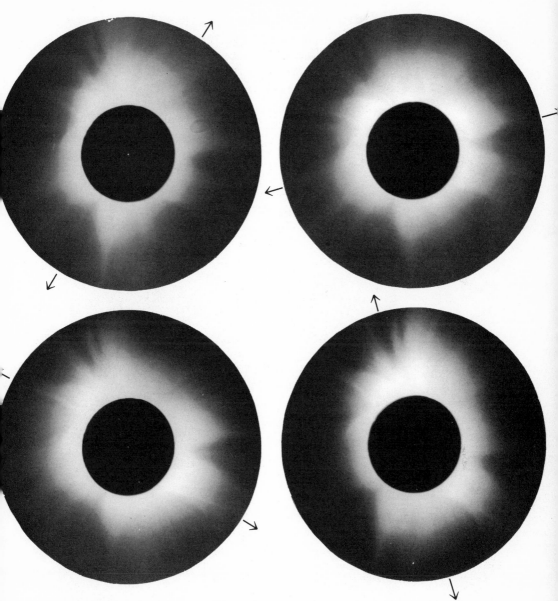

Fig. 124.   Coronal photographs taken through a Polaroid filter, at the eclipse of 8 June 1937. The arrows show the direction of the axis of polarization. Note changes in the coronal form arising from the use of polarized light. (Fernando de Romaña, Peru-Harvard.)

Fig. 125.   B. Edlén, Swedish physicist.

pletely. Edlén specializes in the study of spectra of highly stripped atoms—atoms tortured in a spark until a dozen or more electrons have been torn from them. Starting from a suggestion made by Grotrian, Edlén brilliantly combined experiment and theory to show that highly ionized atoms of iron, nickel, and calcium were responsible for most of the strong coronal lines. Iron, with 9, 10, 12 and 13 electrons stripped from it, produces the strongest coronal emissions. Since neutral iron has 26 electrons, the atom producing the green coronal line (5303 A) has lost exactly half of its electrons. Highly ionized atoms of nickel and calcium account for most of the remaining coronal lines. A few still are of unidentified origin.

The intensities of the emission lines, individually and collectively, vary markedly during the sunspot period. They are, on the whole, most intense near the maximum of solar activity. The green coronal line (5303 A) shows greater range than the red. The red line (6374 A) is the more nearly uniform in distribution around the solar rim. All the lines tend to show greatest brilliance near the spot zones. Thus, in the early part of the cycle, maximum intensity occurs in the middle latitudes. Like the sunspots, with which they show some association, the zones of maximum line excitation drift equatorward as the cycle progresses.

The relative intensities of the various lines show marked changes. Tables giving the brightnesses of the various lines have not been particularly useful. Occasionally they enable us to pick out special relations that help us to understand the physical state of the corona.

At the eclipse of 1932, I noted one special peculiarity. At that time, which was near spot minimum, the green line was fainter than the red. At only one region, a narrow segment of the rim, was the green line at 5303 A at all intense. The sun possessed a green "ear." The red line was also enhanced in this region.

I searched for a related phenomenon and found it. The chromosphere, just below the green area, displayed an enormous intensification of the lines of neutral and ionized helium. Here was the first indication of some connection between chromosphere and corona. Similar records from the 1936 eclipse, where helium was extremely intense near spot maximum confirmed the relation. From this evidence alone I concluded that the then unidentified coronal lines were produced by atoms of high ionization. But all astronomers

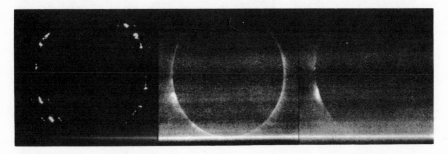

Fig. 126.   Records from 1932 eclipse: (*left*) prominence images in hydrogen (Hα); (*center*) red coronal ring (6374 A); (*right*) green ring (5303 A), showing intensification in one region. (Lick Observatory.)

were, nonetheless, surprised at the denouement, when Edlén showed that the excitation was even higher than they had previously dreamed.

The coronal lines display a character very different from the prominence lines. The latter often show large distortions and displacements, which we attribute to radial velocity (Doppler effect). In comparison, the coronal lines are uniform and show no marked displacements. They are broad, however—a property we ascribe to high temperature.

Usually, the coronal lines appear to avoid regions rich in prominences. In 1947, however, Roberts made a very significant observation with the coronagraph. He noted that the red (6374 A) and yellow (5694 A) lines appeared in the spectrum of an extremely bright prominence. Also these lines were "knotted and twisted" with velocity displacements, like the rest of the prominence lines.

The fact that we can see the corona only around the solar limb is one of the serious limitations of both eclipse and coronagraphic studies. Remember that the corona is three-dimensional and we must attempt to visualize its spatial form.

A study by Roberts of the day-to-day distribution of brightness along the east and west limbs has helped us considerably in this

Fig. 127.   Sections from a spectrogram of 6 January 1956, showing (*left to right*) the green and red coronal lines and Hα from prominences. (High Altitude Observatory, Climax.)

objective. If the corona were constant and rotating with the rest of the sun, a bright patch seen one day on the east limb would show up on the west half a rotation later, and again on the east limb after a full turn of the sun on its axis.

Roberts assumed a variety of rotation periods, ranging from several days to more than a month. He then calculated the correlation between these east and west limb brightnesses for each of these periods. He found a clear maximum of the correlation at an interval of about 13 days. Thus, he concluded (*a*) that the corona was a fairly stable feature and (*b*) that its rotation period was about 26 days, approximately identical with that of the underlying surface.

Plotting the day-to-day records of limb intensities, one can map the bright coronal areas and compare the distributions with the well-known surface features. Figure 128 shows the result of one such attempt. The curves drawn in the figure represent contours of equal coronal brightness. The black areas indicate underlying spot groups. The shaded regions mark the extent of the calcium plages. Note that the bright coronal patches seem roughly centered around

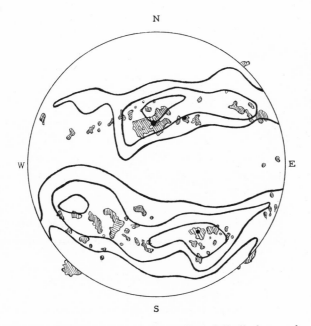

Fig. 128.   Map of the solar corona, built up from daily limb records at Climax. Note the general correspondence of the coronal to the floccular and spot areas; data supplied by McMath-Hulbert Observatory.

the spots and generally confined to the plage areas. The association is doubtless significant but, on the other hand, there is no marked correlation between the brightness of the corona and that of the underlying plage.

⊙

Let us try to assemble the foregoing facts into some sort of theory. The presence of polarization in the corona indicates that the continuous spectrum is scattered sunlight. The blue light of the sky is an analogous phenomenon. Rotate a Polaroid disk and you will see fluctuations in the intensity of the sky radiation transmitted by the Polaroid. This means that the light is polarized.

As for the corona, the primary question is what types of particles are responsible for the scattering. Molecules and very fine dust scatter blue light more effectively than the red. But the corona is white and not blue. Free electrons produce both polarization and white scattered light. However, the electrons are the least massive of all material particles. They move very rapidly in random directions. Thus the combined Doppler shifts from all the electrons in the medium broaden the lines tremendously. I have mentioned that the dark lines of the outer corona are somewhat "washed out" and fuzzy. But the effect to be expected from free electrons is much greater. For the continuous spectrum of the inner corona, where even the strongest dark lines disappear, the hypothesis of electron scattering works very well. Most astronomers have, tentatively at least, agreed that electrons cause the inner coronal light. The temperature necessary to make the lines disappear is the enormous value of 1,000,000°C.

Free protons, or hydrogen nuclei, also scatter radiation, but only about 1/4,000,000 as effectively as electrons. Since the gas must contain electrons and protons in nearly equal numbers, we cannot possibly ascribe the spectrum to the action of protons.

Van de Hulst and Allen independently have suggested that dust between the earth and the sun is responsible for that portion of coronal light exhibiting a solar absorption spectrum. This type of scattering should produce a symmetrical halo around the sun, analogous to that formed by dust in the earth's atmosphere, except that the latter is ordinarily much the brighter. This suggestion would make part of the coronal light a sort of haze from interplanetary space.

Small particles, somewhat larger than the wavelength of light, would scatter light in almost the observed fashion. Here the difficulty is to explain how the bits of solid matter can condense from a gas as hot as the corona, especially in the environs of the sun. Or are they particles of meteoric dust, falling slowly into the solar furnace?

They cannot be permanent members of the solar atmosphere. The changing form of the corona requires that they be continuously replenished. The existence of the zodiacal light, perhaps a sort of extension of the corona in the plane of the planetary orbits, may be regarded as a partial confirmation of the meteoric-dust hypothesis. You can see this light—a hazy patch extending upward along the ecliptic—before sunrise in autumn or after sunset in the spring, in the northern hemisphere.

The actual structure of the corona indicates that its most brilliant portions, however, must be a true solar phenomenon and attributed to a combination of electron scattering and bright emission lines. The outer regions, beyond a solar radius or so, may, in part, be a sort of zodiacal-light phenomenon. How one can explain the variations of form of the outer corona in terms of the sunspot cycle is not yet clear. Perhaps we shall find, after all, some atomic process as yet unrecognized, that can produce a scattered solar spectrum.

The interpretation of the bright-line spectrum leads us into further difficulties. We have already seen that a minimum temperature of $20,000°\,K$ is required to explain the chromospheric and prominence spectrum, where helium has lost a single electron. The total energy necessary to strip 13 electrons from an atom of iron is far greater. Calculations show that we must postulate temperatures in excess of $1,000,000°\,K$ to explain the observed stripping.

The requirement of a million-degree temperature, moreover, appears in at least six independent analyses. The observed excitation in the emission lines and the complete disappearance of the scattered Fraunhofer lines are two cases already cited. The breadths of the emission lines in the corona are consistent with the interpretation, as a Doppler effect, of iron atoms moving in accord with a temperature $1,000,000°$. Only at a temperature of the order of $1,000,000°$ will the emission from abundant hydrogen be effectively quenched. And, if we interpret the corona as an atmosphere of ionized hydrogen in equilibrium against the expansive force of high temperature and the contractive force of gravitation, the derived temperature again comes out about 1,000,000°. The sixth line of evidence, from radio astronomy, will be discussed in the next section.

In regions of specially high excitation, for example near surges or loop prominences, Roberts has found that the coronal lines tend to be even wider than in ordinary regions. The temperatures calculated from such observations may reach as much as 5,000,000°. Moreover, the yellow lines attributed to Ca XV are wider in just the amount consistent with the fact that calcium atoms are lighter and therefore faster moving than those of iron. The excess heating in such regions probably comes from shock waves in gas ejected from the active spot areas.

Fig. 129.   Green-line corona taken with coronagraph and filter. (Sacramento Peak Observatory.)

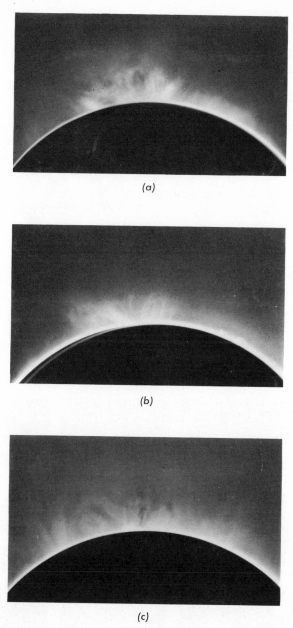

(a)

(b)

(c)

Fig. 130. Photographs of the solar corona, taken without eclipse, with the aid of a coronagraph and monochromatic filters: (*a*) 8^{h}00^m 3 September 1941, 5303 A; (*b*) 8^{h}15^m 3 September 1941, 6374 A; (*c*) 9^{h}00^m 14 September 1941, 5303 A. Note the marked structural differences in the green (*a*) and red (*b*) images, taken within a few minutes of each other. (Lyot, Pic du Midi.)

Chapman has suggested that heat from the solar corona leaks slowly outward primarily by conduction rather than by radiation or convection. In such a corona, extending far beyond the earth, the temperature gradually declines from its million-degree value near the sun to about 200,000° at the distance of the earth. It is startling to imagine that, only a few thousand miles above our heads, probably lies a tremendously hot region of space. This high temperature would have no effect on man or on space vehicles traveling through the region, because the density of particles is far too low. Some heat, conducted into the upper earth's atmosphere, may, however, influence conditions in the ionosphere.

⊙

Radio Noise
from the Sun   Within recent years a new and completely independent line of evidence on conditions in the corona has come to hand. The story began dramatically with the use of radar during World War II. One afternoon in 1942, British "early-warning" radars went berserk. Strong high-frequency radio waves blanketed the regular radar signal. At first, the operators suspected a new enemy counter-

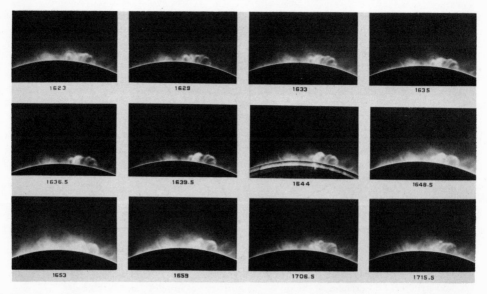

Fig. 131.   Solar corona in light of 5303 A, 22 November 1956. (Sacramento Peak Observatory.)

measure. But a check showed that all the radars on the coast were pointing toward the setting sun—on whose disk at the time was a giant sunspot.

Further observations, during and soon after the war, established that the sun was indeed a source of high-frequency radio emission. From study of these emissions a whole new branch of solar astronomy has developed.

Radio waves are, of course, electromagnetic radiations (Chapter 3), the same as visible light except of enormously greater wavelength. The observable radio spectrum extends from about 1 centimeter to 10 meters. Like the optical spectrum, the radio spectrum is limited on its short-wavelength end by absorption in the earth's atmosphere, in this case by molecules of oxygen and water vapor. On the long-wavelength end our lower atmosphere is always transparent, even on cloudy days. But a high layer, called the ionosphere (see Chapter 14), begins to interfere at around 10 meters. The complete electromagnetic spectrum of the sun, together with the atmospheric absorption, is illustrated schematically in Fig. 132.

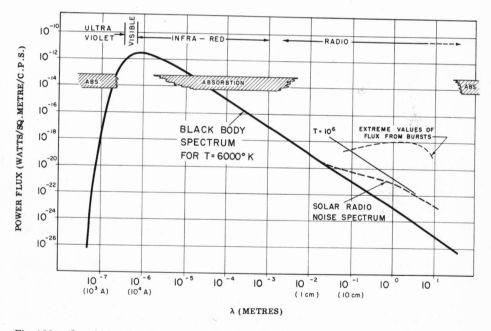

Fig. 132.   Graph showing the energy radiated by the sun at different wavelengths. (Covington, National Research Council, Canada.)

In Chapter 4, I indicated that the energy distribution in the visible spectrum corresponds to a temperature of 5800°K. As you can see from Fig. 132, this is not true for the radio emission, which seems to require a different temperature for each wavelength. In the band from about 10 meters to 1 meter (30 to 300 megacycles per second) the intensity of the radiation is enormously greater than the expected "heat energy" in these wavelengths. The indicated temperatures increase systematically with wavelength, from values below 10,000°K at 1 centimeter (30,000 megacycles per second) to 1,000,000° at the meter wavelengths. This at first surprising fact actually has a simple explanation in terms of the nature of radio waves and the structure of the solar atmosphere.

The radio waves are radiated by fast-moving electrons in the highly ionized gases of the outer solar atmosphere. Ionized gases which are fully transparent to visible light, however, may be opaque to radio waves of certain wavelengths. The opacity depends on the density of the ionized gas, or more precisely on the number of free electrons per cubic centimeter. In the chromosphere, where the density is relatively high, the gases are completely opaque to meter wavelengths; only the centimeter waves can escape to reach the earth. Meter waves can escape only from the more tenuous corona, not from the lower and denser layers.

You can readily see that this fact should cause the solar disk to appear larger in radio than in optical light. Moreover, the apparent diameter is directly related to the wavelength of the radio emission. The meter-wavelength sun is larger than the centimeter-wavelength sun, and both are larger than the visible disk.

The longer the wavelength, the higher the level in the solar atmosphere the waves come from. This property has given astronomers a powerful new tool for studying the structure of the chromosphere and corona. The observed minimum temperature of 1,000,000° in the meter wavelengths is directly related to the kinetic energy of the free electrons in the corona, and provides the sixth line of evidence for such high coronal temperatures.

Note that I said *minimum value* of 1,000,000°. Actually the amount of energy radiated in these frequencies shows wide fluctuations. During a violent outburst of radiation in the radio range of wavelengths, observed as radio noise, we should require temperatures in excess of 1,000,000,000° to account for the emission as a temperature phenomenon. Probably, however, other more complex proc-

esses involving shock waves and oscillations of ionized gases contribute to these intense outbursts. Scientists early observed that
most of this enhanced radiation came from limited areas of the
disk, from the neighborhood of active sunspots.

The variability of solar radio noise depends strongly on wavelength. For waves shorter than 2–3 centimeters, the intensity is
steady and undisturbed. From 3 to 60 centimeters or so, often called
the decimeter range, the intensity shows occasional short-lived increases or *bursts*. These tend to last for a few minutes and are often
associated with solar flares. The decimeter intensity also shows a
slowly varying component which tends to exhibit a 27-day period
associated with the solar rotation, and arises from the vicinity of
active spot regions. This component also varies systematically with
the sunspot cycle (see Fig. 133).

As you can see from Fig. 134, the base level for meter-wavelength
radiation is again relatively constant. But intense and rapidly varying disturbances occur above this level. Large sporadic *outbursts*,
lasting for minutes, occur often in association with the bigger solar
flares. A million-fold increase in intensity within a few seconds has

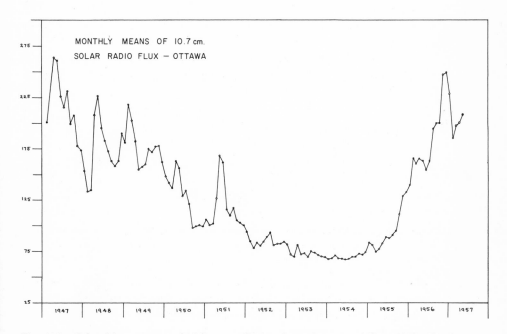

Fig. 133.   Monthly averages of 10.7-cm radiation from the sun, 1947 to 1957. (Covington,
National Research Council, Canada.)

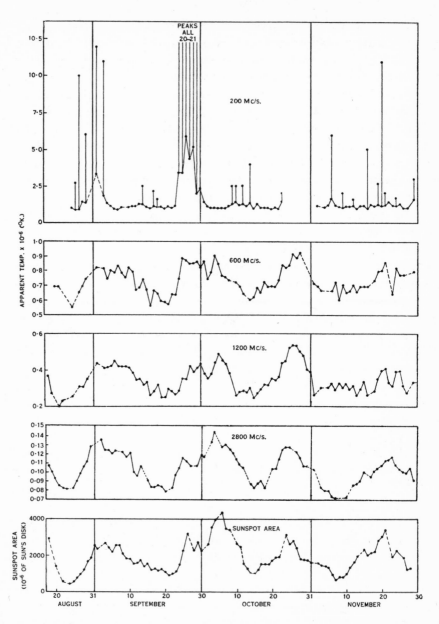

Fig. 134.   The variation with time of the apparent disk temperature at frequencies of 200, 600, 1200, and 2800 Mc/s, together with the projected sunspot area for the period, August—November 1947. (Piddington and Minnett.)

been observed in the most violent outbursts. Some large and active spot regions also produce *noise storms,* a long series of bursts or a rise in the base level, lasting for hours or even days. Denisse, a French radio astronomer, has found that these "noisy spots" are likely to spray the earth with corpuscular radiation, to produce the northern lights and magnetic disturbances which I shall describe in Chapter 14. "Quiet spots," on the other hand, generally move across the solar disk without troubling us here on earth.

Australian radio astronomers first pointed out that bursts on the higher frequencies arrive slightly earlier than those on the lower frequencies. They interpreted these delays as evidence of a disturbance moving outward through the solar atmosphere, at a velocity of about 1000 kilometers per second, and passing through regions from which successively lower frequencies could escape. The calculated velocity of 1000 km/sec is similar to that indicated for the ionized clouds coming from the sun and causing aurorae and magnetic disturbances here on earth. Various scientists have suggested that streams of particles, sometimes visible as surges or jets, passing out through the corona produce these disturbances. While this interpretation appears straightforward and attractive in some respects, other scientists have attributed the disturbance to a shock wave passing through the solar atmosphere. The field is still too new for us to choose with any certainty between these explanations.

The most complete analysis of time-delay phenomena has been made by J. P. Wild, whose "Dynamic Spectrum Analyzer" is a sort of spectrograph for radio waves. A simple receiver can measure the intensity of radio noise over only a narrow band of frequencies. The spectrum analyzer, however, can be rapidly and automatically tuned over a wide range and will register the noise intensity over this band. In addition to the so-called slow bursts which I described in the preceding paragraph, Wild has observed others which he calls fast bursts. These have a much shorter time delay and are attributed to disturbances moving outward at 30 to 50 thousand kilometers per second.

Harvard has recently installed a spectrum analyzer near Fort Davis, Texas. With this instrument, A. Maxwell and co-workers have observed fast bursts, slow bursts, and a remarkable new type which they call the "inverted-U burst" (Fig. 138). The inverted-U

Fig. 135.   Fort Davis solar radio station. Antenna by Jasik; receivers by Airborne Instruments Laboratory; built and operated by Harvard College Observatory, under contract with Air Force Cambridge Research Center.

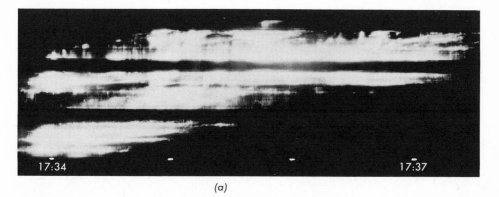

(a)

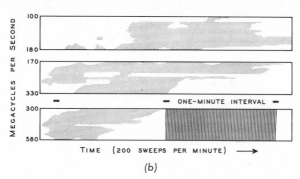

(b)

Fig. 136.   A slow-drift burst of radio noise of great intensity, from 17ʰ34ᵐ to 17ʰ37ᵐ U. T. 7 January 1957: (*a*) actual record, from Fort Davis; (*b*) schematic diagram. Each of the three strips depicts the intensity in a given frequency range as indicated. Note that the disturbance begins at the higher frequencies and drifts toward the lower ones. (*Sky and Telescope.*)

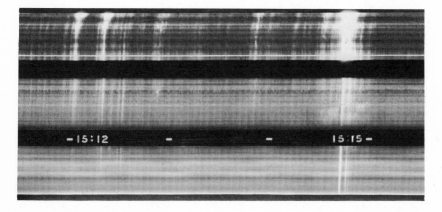

Fig. 137.   Fast-drift bursts, recorded on 4 February 1958. (Fort Davis.)

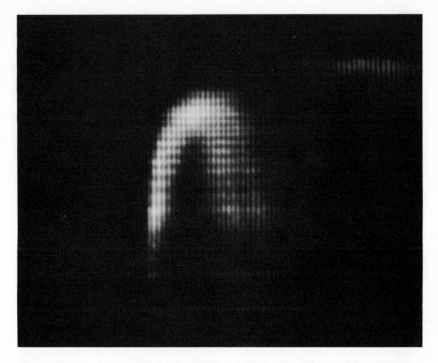

Fig. 138. An inverted-U burst, 29 November 1956, a form of fast-drift burst which starts at high frequencies, drifts to lower, and then returns to higher. (Fort Davis.)

burst suggests a disturbance moving upward and then returning to lower levels. Krook has proposed that these disturbances may be shock waves traveling along the magnetic lines of force between oppositely polarized members of a bipolar spot group.

A different type of refinement on the simple receiver has been made by W. N. Christiansen and co-workers of the Australian Radiophysics Laboratory. Unlike optical telescopes, radio receivers have a poor resolving power, which becomes worse the longer the wavelength. Even gross details of structure cannot be observed without special equipment utilizing the principle of wave interference (see p. 44) or interferometry. Christiansen has used two independent antenna systems, one having high resolution in the E-W plane, the other in the N-S plane, to obtain the distribution of intensity over the solar disk, shown in Fig. 139. You can see that the 21-centimeter sun is somewhat flattened and ellipsoidal, and that the regions of greatest intensity lie in the sunspot zone, near the edge of the optical disk. This effect is called *limb brightening* and is analogous to limb darkening (see p. 157), only in reverse because the 21-centimeter radiation arises from higher and hotter levels of the atmosphere.

Gold and Menzel have suggested that the explosive activity of the prominences and inner corona might produce electromagnetic waves of extremely long wavelength, whose frequencies were in the subaudible range of about 1 cycle per second. Indeed, some years ago, Menzel and Salisbury had found evidence for the existence of such waves. The outward streaming of the gaseous envelope tends to "comb" the magnetic field of the corona into roughly radial form. This radial magnetic field would tend to guide the waves earthward and the shorter waves caused by an individual blast would tend to arrive earlier than the longer ones.

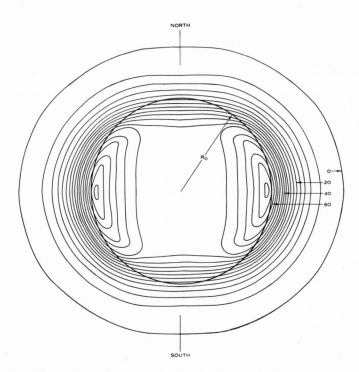

Fig. 139.   Derived two-dimensional radio brightness distribution at the wavelength of 21 cm for the quiet sun. The contours are at equal intervals of 4000°K. The central brightness temperature is 47,000°K, and the maximum peak brightness is 68,000°K. (Christiansen and Warburton.)

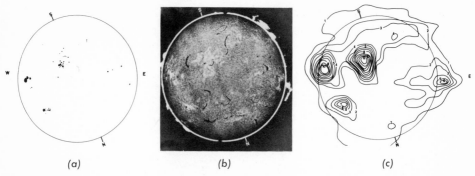

(a)   (b)   (c)

Fig. 140.   Pictures of the sun on 4 December 1957: (*a*) in white light; (*b*) in Hα (Division of Physics, C.S.I.R.O., Sydney); (*c*) in emission of 20-cm radiation (Division of Radiophysics, C.S.I.R.O., Sydney). The radio emissions appear at heights of 40,000 to 100,000 km above chromospheric plages. The radio emitting source outside the east limb is related to an active region which has not yet appeared on the photospheric and chromospheric pictures. It does not coincide in position with the prominence in the Hα photograph. The contour intervals in the radio picture correspond to brightness differences of 130,000°K.

A somewhat similar effect occurs in the earth's atmosphere. A lightning flash produces a sharp pulse of audible radio frequencies, which follow north and south along the earth's magnetic field to the opposite hemisphere. The higher frequencies arrive first so that the impulse sounds like a whistle of descending pitch. The noise records obtained by C. S. Wright of Canada show a similar effect —with far lower frequencies. Gold and Menzel term the phenomenon *solar whistlers,* on the supposition that they are actually of solar origin.

⊙

Problems of Coronal Temperature   The high temperatures of the corona, prominences, and chromosphere present many problems. How can one explain them in terms of the relatively feeble 5800° temperature of the visible surface layers? The second law of thermodynamics—a basic physical principle—states that heat always flows from a hot body to a cool one, never in the reverse direction. One does not put the teakettle in the refrigerator for the purpose of boiling water. There seems to be no obvious way that the surface temperature can produce a higher temperature in the upper regions.

One is inclined to turn to the hotter depths of the sun for an explanation. The temperature rapidly increases in the subsurface regions toward the value of about 15,000,000° at the center. If a sunquake or solar volcano could force the hot material to the surface, we should have a ready supply of energy. The sun is gaseous throughout. Can ascending currents elevate masses of heated gas from the interior? Or can sunspots, by vortical motion, act as centrifugal pumps to raise the hot matter?

At one time, I definitely favored theories based on such principles. There are difficulties, however. Suppose we were to encase a portion of the hot solar gas in a weightless heatproof envelope. Also suppose that we were to warm the gas slightly, so that the material expanded and thus became lighter than its surroundings. The bubble would then act as a balloon and slowly rise toward the surface.

With the aid of mathematics and physics, we can trace the progress of the ascension. The balloon will slowly expand and the gases will cool as a consequence. We find, however, that the cooling is greater than that of the surrounding atmosphere. At some point, the gas will become denser than the surroundings and the balloon will start to descend. Thus the bubble will alternately rise and fall until friction damps out the oscillations. It will never reach the surface.

The fictitious weightless envelope of the gaseous bubble has no real effect on the problem, except to hold the original gas together. No analysis that I have made suggests that we could get the heated gases out from the deep interior. Or, if we should resort to force and pump them out, they would be cooler than the visible surface. We see in this discussion a possible explanation of sunspots, but none for the high excitations of prominences and corona.

We have already noted that a turbulent layer probably occurs near the surface of the sun, in a region where the temperature ranges from 10,000° to 20,000°C. This zone of convection arises from ionization of hydrogen. Most astronomers agree that the region is a likely source of the turbulence associated with granules and spicules. Thomas and Houtgast have suggested that jets of gas shoot out with supersonic velocities and cause strong shock waves, whose energy heats the corona to the observed high temperatures. M. Schwarzschild has recently made a similar suggestion, except that his waves or pulses are subsonic. It is highly probable that effects of this type play an important role in determining the character of the solar corona, as well as that of the chromosphere, as detailed in Chapter 8.

Fig. 141.   Corona, 3 January 1908.
(Lick Observatory.)

Fig. 142.   Corona, 8 June 1918.
(Mount Wilson Observatory.)

Fig. 143.   Corona, 21 September 1922.
(Lick Observatory.)

Fig. 144.   Corona, 10 September 1923.
(Sproul Observatory.)

Fig. 146.   Corona, 21 October 1930.
(Sproul Observatory.)

Fig. 145.   Corona, 14 January 1926.

Fig. 147.
Corona, 31 August 1932.
(Wright, Lick
Observatory.)

Fig. 148.   Corona, 8 June 1937, specially
printed to show inner and outer structure.
(Gardner; copyright, National Geographic
Society.)

Fig. 150.   Corona, 30 June 1954.
            (Roland Rustad, Jr., Minneapolis.)

Fig. 149
Corona,
25 February 1952.
(Van Biesbroeck.)

Alternative recent proposals have involved release of atomic energy. In Chapter 11, we shall discuss some problems concerning nuclear energy. For the present, I shall not go into detail. Saha postulates the existence of nuclear fissions similar to those made famous by the atomic bomb, but employing different atoms as agents. Woolley is less revolutionary in his suggestion that gases ascending from the interior may possess radioactive products of the so-called *carbon cycle,* one of the atomic processes that might be responsible for energy generation in the sun and stars (see Goldberg and Aller, *Atoms, Stars, and Nebulae;* see also Chapter 11).

Both of those suggestions suffer from the same fundamental defect—lack of quantitative agreement. The radiations from the chromosphere and corona comprise something less than 1 percent of the total radiation of the sun. But the main body of solar energy must originate in a core whose total mass is roughly one-tenth that of the sun. The chromosphere and corona together certainly comprise less than a million-millionth part of the solar mass. Even if we required them to furnish only a hundred-thousandth of the total radiation, we should need gases capable of producing energy one million times more efficiently than those in the deep interior. I can see no escape from these figures. We must discard nuclear processes as a form of excitation.

There is another possibility, which may be discussed in several forms. They all depend on the presence of variable magnetic fields and, at least by implication, also variable electric currents in the sun. Alfvén and Walén have postulated the existence of *hydrodynamic electromagnetic waves,* flowing slowly outward from the depths (see p. 126). The concept is interesting, and further development of the principle may shed light on many solar problems. At present, the hypothesis is too new for detailed treatment, though I shall refer to some of its features in a later chapter.

As any student of physics knows, every electric current possesses an associated magnetic field. Also, every *changing* magnetic field will give rise to an electric field. Note the emphasis on *changing.* A constant field will produce no current whatever. A conductor moving through an otherwise static magnetic field causes changes in that field which in turn induce an electric field and currents in the conductor. The mathematical relations between electricity and magnetism are well known. Maxwell formulated the laws in 1873. All electrical and radio engineers use these equations as part of their fundamental tools.

The coronal lines are completely invisible in the ordinary solar spectrum. They have been observed as bright lines in the spectra of several exploding stars, the so-called *repeating novae*—stars that have shown more than one outburst. The association of these lines with such stellar cataclysms is significant. In no way are we to infer that the sun is subject to such a disaster. Nevertheless, we conclude that only the highest temperatures are great enough for the wholesale stripping of atoms necessary to produce the coronal lines.

In science, we face what seems to be a never-ending series of problems. The solution of one raises new difficulties. The quest for knowledge goes on. In the relative sense, every day increases our ignorance.

10

Solar Eclipses—Old and New

Of all astronomical phenomena, total solar eclipses are the most spectacular. The darkened sky, the hushed atmosphere, the on-rushing shadow, the narrowing crescent, the "Baily's beads" formed where the sunlight filters between adjacent mountains of the moon, the silvery halo of the corona outlining the black lunar disk—small wonder that the sight has inspired awe and fear throughout the ages!

Why do eclipses differ from one another? And what methods does the astronomer use in forecasting the circumstances of an eclipse? Solar eclipses occur, as is well known, when the moon comes into direct line between the sun and the earth. The apparent paths of the moon and the sun in the sky are circles, the moon making twelve and a fraction complete circuits while the sun is making only one, in the course of a year. Unfortunately for the astronomer, these two paths are inclined to one another, so that the moon crosses the sun's track twice each month. I say "unfortunately," because if the two

paths coincided exactly, we should have one solar eclipse each month. Every new moon would cut across the solar disk. The astronomer would welcome such frequent opportunities for making eclipse observations.

But nature has decreed otherwise and rectification of this fault in the design of the cosmic machinery is beyond our power. Eclipses will occur only when both bodies are simultaneously near the *nodes*, as the two points of intersection of the orbits are called. If the sun

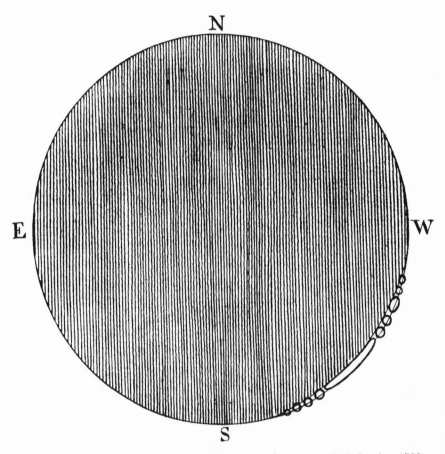

Fig. 151.   Earliest record of "Baily's Beads." At the eclipse of 27 October 1780, during the Revolutionary War, Harvard and the American Academy of Arts and Sciences sent an expedition to Penobscot Bay. Probably because of an error in the tables, the party located outside rather than in the totality belt. At that time Williams observed the beads, and made the drawing reproduced above, 56 years before Baily made his studies.

crosses a node, say on 1 January an eclipse will occur within two weeks of that time, when the moon happens to come along. But by 1 February the sun will have moved away from the intersection, the new moon will pass either above or below the sun, and there will be no chance for an eclipse until about six months later, when the sun reaches the opposite node. Eclipses thus occur in two seasons, spaced approximately half a year apart.

To simplify the argument, imagine the sun and moon circling a stationary earth. We may then aptly compare the sun and moon to two racers, one of whom (the moon) is mounted on a bicycle and therefore able to travel much faster than the other (the sun). At the start of the race they are lined up in front of the judges' stand (the node). I shall use this alignment of moon, sun, and node to represent an eclipse. The starting gun cracks. The cosmic race is on! The moon dashes away and the sun trots leisurely around the

Fig. 152.   Composite picture schematically illustrating the conditions of a total solar eclipse. At an actual eclipse the tapering lunar shadow may be anywhere from 0 to 100 miles wide where it falls upon the earth. The background is an infrared photograph taken from a Viking 12 rocket at an altitude of 143 miles. Part of Mexico, the Gulf of California, and lower California, extending up to Los Angeles are clearly visible. To the right of the Gulf of California are the Laguna Salada and Salton Sea. In the lower right appears the area of Phoenix, Arizona. (Official U. S. Navy Photograph.)

celestial track. When the moon passes the judges' stand after one circuit the sun will be at some entirely different point of the course so that the circumstances that gave rise to the initial eclipse do not occur. They do not pass one another at the judges' stand.

The race continues. The sun and moon make numerous circuits of the track. Finally, as the never weary sun passes the stand after his nineteenth circuit, along comes the moon, whizzing by just as the sun crosses the starting line. The original alignment of moon, sun, and node recurs, and the eclipse repeats under conditions very closely approximating those at the beginning of the race.

We meet one further complication on the cosmic track, not found on terrestrial ones. The judges' stand is not stationary. The lunar orbit itself is gradually shifting its position and the point of intersection with the sun's apparent path changes in consequence. The effect is a slow backward motion of the node. We must imagine the judges' stand to be on wheels and creeping around the cosmic track in a direction counter to that of the racers. Thus the sun passes the stand at intervals of 346.62 days instead of 365.24. The former figure is often called an eclipse year.

⊙

Forecasting of Eclipses

The average interval between successive new moons is 29.53059 days. Nineteen eclipse years are equal to 6585.78 days. Also 223 lunations (the time from one new moon to the next) take 6585.32 days, corresponding to 18 years 11⅓ days (or 10⅓ days if five leap years have intervened). We thus check numerically the statement of the preceding paragraph that, after nineteen eclipse years, the conditions that gave rise to the original eclipse are very nearly repeated. Of course other eclipses have taken place during the interval, but they occur at different distances from the exact nodes and hence bear no simple relation to those we are considering. One might say that there are other judges' stands not exactly on the line of nodes.

· The interval of 18 years 11⅓ days, which is called the *saros*, was discovered late in the 17th century by Edmund Halley. As an example, a total eclipse visible from Labrador, Spain, and Egypt occurred on 30 August 1905. Adding 18 years and 11 days, we find that a related eclipse should have occurred on 10 September 1923. We must not forget the extra one-third of a day, however. For during those 8 hours, the earth made one-third of a revolution. Consequently the 1923 track lay 120° (⅓ of 360°) to the west of the 1905 path and crossed California and Mexico. This eclipse is remembered more for prevailing clouds and widespread disappointment than for scientific results.

We must regard the saros as an accidental interval, although the shifting of the lunar orbit is controlled, of course, by definite gravitational forces, such as the pull of the sun and the effect of an earth that is not exactly spherical. (There is no true accident in solar-system mechanics!) But a second not-so-well-known feature of the saros makes the relation all the more remarkable.

The character of an eclipse depends markedly upon the distance of the moon from the earth. This distance varies because the moon's orbit is an ellipse, not a circle. When the moon is far away its apparent diameter will be less than that of the sun and the eclipse will be annular; a ring of sunlight will circle the black lunar disk. When the moon is near the earth its apparent diameter will exceed that of the sun, and the eclipse will be total. The position of the longer axis of the ellipse does not remain fixed in space. But, by a remarkable coincidence, 18 years 11½ days serve to bring the orbit back into its original orientation. Thus after the lapse of one saros we not only find the sun and moon back in position at the node, but also discover that the moon is at very nearly its original distance from the earth. Successive eclipses after a saros are hence very similar in character, which they would not be if this second accidental relation did not exist.

The moon possesses a long tapering shadow pointing away from the sun. The eclipse path, to which I have previously referred, is merely the track of this shadow as it crosses the earth. It represents the zone for which the eclipse is total. If an eclipse occurs when the moon is in the outer portion of its orbit, the tip of the shadow will fall short of the earth. Under these circumstances, the eclipse will be *annular,* and the "path" will then represent the region from which the moon will appear to lie within the solar disk. An observer outside the path will, in either case, find the sun incompletely covered, and thus will view a partial eclipse.

Other periods are known in addition to that of the saros. One is equal to 29 years, less 20.3 days. Eighteen of these 29-year intervals are equal to 521 years, minus a small fraction of a day. Successive eclipses of the 29-year cycle occur in different portions of the sky and the shadow paths hit entirely different regions of the earth.

Fig. 153.   Various aspects of the annular eclipse of 7 April 1940.

Two of the 29-year cycles (58 years) are very roughly equal to three of the 18-year periods (54 years). The elliptic orbit of the moon rotates about 1½ times in 29 years. Hence, if one eclipse of the 29-year series is total, the next will very likely be annular, the third total, and so on. This alternation of types, however, is not exact. The total eclipse of 21 September 1903 was followed by the eclipse of 31 August 1932, also total. But the next eclipse of the cycle, 11 August 1961, will be annular. That of 22 July 1990 will be total.

Fig. 154.   Partially eclipsed sun, 12 November 1947. (Fairmount Solar Station.)

To aid the student who may wish to check on the occurrence of eclipses—past, present, and future—I give in Table 5 a list of all total, annular, and partial eclipses occurring in the twentieth century. The eclipses are grouped, in order of occurrence, according to each individual saros epoch. The successive columns give the day, month, and year of the eclipse, the type, and the latitudes and longitudes at the beginning, middle, and end of the track. Minus signs denote east and plus signs west longitude. The last column gives the maximum duration of the total or annular phase in minutes and seconds. Equatorial eclipses are generally of longer duration than polar ones because of the earth's rotation, which causes the observer to chase along in the direction of the shadow drift.

Both total and annular eclipses appear as partial eclipses outside the belt of totality. Thus the table of eclipses gives as partial those that occur near the poles, where the shadow point completely misses the earth.

To illustrate still further the characteristics of the saros relation, let us follow the evolution of the first set of eclipses through the century, from 28 May 1900 to 22 July 1990. Each successive eclipse is farther north than the preceding one. The duration changes but slightly. This series of eclipses began on 10 March 1179 (Julian calendar) as a partial eclipse near the south pole, and formed a long series of annular eclipses from 1323 to 1828. Gradually, the inequalities in the varying lunar distance changed the type to total, the first of which came in 1864. In 2062, the eclipses will become partial again. The final performance of the series, as a partial eclipse near the north pole, will take place about A.D. 2200. Thus, we cannot project the saros cycle indefinitely into the future. For long ranges, the 521-year cycle is better, although it gives no certain indication of the type of eclipse or of its location on the surface of the earth.

During the twentieth century, three series have drifted from total to partial phases. During the same interval, three new series of total eclipses have come in. Some of the tracks show a drift to the south, whereas an approximately equal number drift northward. Usually, tracks of totality for eclipses occurring at a given node will tend to drift in the same direction. There are, however, some exceptions to this rule.

In passing, I wish to call special attention to the unusual eclipses marked C (central). In these the point of the tapering shadow barely reaches the earth. Theoretically, the totality lasts for a second or two, but practically the eclipse is annular. At maximum obscuration, the photosphere shines through valleys or other depressions of the lunar surface. Baily's beads form a complete necklace, spoiling the total phase and concealing the corona by their brilliance.

From the data given in Table 5, anyone who wishes may determine the path of a given eclipse. First plot the three points, denoting beginning, middle, and end of totality, on a map of the world. Do not use a Mercator projection, in which the lines are all greatly distorted. Instead, employ either a globe or a flat map drawn on the *stereographic* projection. Through the three points, describe an arc of a circle. This arc will represent the approximate path of the total or annular phase. But the data are not sufficiently accurate to determine whether the eclipse will be truly total or only nearly total at any given location.

The most complete series of approximate eclipse predictions is due to Oppolzer, from whose catalogue I have taken some of the data of Table 5. His computations cover the complete range from 10 November 1207 B.C. to 17 November A.D. 2161.

The detailed calculation of all the circumstances of an eclipse— the precise position of the path, the exact moments when eclipse begins and ends, and so on—is clearly a matter for technical mathematics. Ten years or so in advance of the event, the astronomer

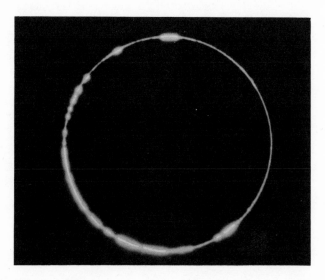

Fig. 155.   The "total" or central eclipse of 28 April 1930. (Lick Observatory.)

Table 5.   ECLIPSES FROM A.D. 1900 TO A.D. 2000, ARRANGED BY SAROS.[a]

Date	Type[b]	Beginning		"Noon"		End		Maximum Duration
		Latitude	Longitude	Latitude	Longitude	Latitude	Longitude	
May 28 1900	T	+17° 50.3	+116° 38.4	+44° 56.8	+45° 0.4	+25° 20.6	−31° 37.1	2^m 8.8^s
June 8 1918	T	+25 41	−129 58	+50 51	+152 10	+25 23	−74 31	2 23.0
June 19 1936	T	+33 51	−15 58	+56 24	−101 26	+25 36	−179 37	2 31.7
June 30 1954	T	+42 22	+99 4	+61 53	+4 32	+26 19	−73 57	2 35.0
July 10 1972	T	+51	−144	+67	+111	+28	+32	3 —
July 22 1990	T	+60	−24	+73	−142	+30	+139	3 —
Nov. 21 1900	A	+5° 59.7	−2° 41.1	−33° 19.1	−65° 49.9	−18° 26.8	−135° 19.4	6^m 43.7^s
Dec. 3 1918	A	−10 36	+119 7	−36 5	+53 19	−15 4	−14 59	7 6.0
Dec. 14 1936	A	−15 21	−118 12	−37 51	+173 1	−11 14	+106 40	7 25.6
Dec. 25 1954	A	−20 11	+4 58	−38 40	−66 52	−7 13	−130 49	7 39.6
Jan. 4 1973	A	−25	+128	−39	+54	−3	+8	8 —
Jan. 15 1991	A	−30	−109	−38	+174	0	+114	9 —
May 17 1901	T	−27° 27.6	−40° 11.2	−2° 7.1	−96° 51.9	−12° 49.0	−156° 53.6	6^m 26.8^s
May 29 1919	T	−19 43	+75 9	+4 18	+17 23	−12 25	−42 27	6 50.6
June 8 1937	T	−11 47	−169 44	+9 54	+130 27	−12 23	+70 51	7 4.0
June 20 1955	T	−3 42	−54 49	+14 41	−117 26	−12 34	−176 47	7 7.8
June 30 1973	T	+5	+60	+19	−6	−13	−65	7 —
July 11 1991	T	+13	+175	+22	+105	−13	+46	6 —
Nov. 10 1901	A	+36° 55.7	−13° 34.0	+11° 44.6	−66° 29.9	+17° 20.8	−122° 7.9	11^m 2.5^s
Nov. 22 1919	A	+31 41	+102 31	+7 18	+50 24	+19 11	−4 11	11 36.9
Dec. 3 1937	A	+26 22	−139 25	+4 4	+168 21	+21 47	+115 0	12 0.0
Dec. 14 1955	A	+21 19	−19 40	+2 7	−72 39	+25 10	−124 36	12 9.3
Dec. 24 1973	A	+16	+102	+1	+47	+29	−3	12 —
Jan. 4 1992	A	+11	−137	+2	+168	+33	+118	12 —
April 8 1902	P	Arctic	Last Eclipse of Saros					

Date	Type							Duration
7 May 1902	P	Antarctic						
17 May 1920	P	Antarctic						
29 May 1938	T	−65° 40′	+50° 21′	−53° 27′	+26° 29′	−61° 20′	−8° 33′	$4^m\ 4^s.0$
8 June 1956	T	−55 25	−179 2	−40 47	+140 26	−55 59	+100 32	$4\ 44.7$
20 June 1974	T	−45	−59	−32	−107	−53	−149	—
30 June 1992	T	−35	−35	−26	+5	−51	−39	—
30 Oct. 1902	P	Arctic						
10 Nov. 1920	P	Arctic						
22 Nov. 1938	P	Arctic						
2 Dec. 1956	P	Arctic						
13 Dec. 1974	P	Arctic						
24 Dec. 1992	P	Arctic						
28 Mar. 1903	A	+39° 52.3	−80° 8.1	+65° 13.4	−150° 0.1	+74° 56.4	+116° 45.6	$1^m\ 56^s.7$
7 Apr. 1921	A	+45 41	+42 38	+75 28	−34 19	+77 30	−153 5	$1\ 54.4$
19 Apr. 1939	A	+53 41	+167 22	+88 23	+78 42	+78 6	−76 42	$1\ 52.6$
29 Apr. 1957	A	+65 34	−56 30	+70 40	−40 4	+74 13	−17 17	
11 May 1975	P	Arctic						
21 May 1993	P	Arctic						
20 Sep. 1903	T	−46° 17.2	−31° 13.4	−69° 55.2	−100° 46.4	−82° 0.3	−178° 32.3	$2^m\ 14^s.9$
1 Oct. 1921	T	−52 28	+96 33	−84 23	+19 23	−86 20	−127 14	$1\ 52.0$
12 Oct. 1939	T	−60 0	−129 34	−72	−165	−81 28	−72 33	$1\ 32.5$
23 Oct. 1957	T	−69 31	+18 37	−71 15	+23 20	−72 45	+28 13	—
3 Nov. 1975	P	Antarctic						
13 Nov. 1993	P	Antarctic						
16 Mar. 1904	A	−10° 15.3	−35° 53.6	+6° 20.6	−95° 44.8	+25° 12.4	−157° 3.7	$8^m\ 2^s.4$
28 Mar. 1922	A	−7 43	+75 32	+13 14	+16 34	+27 29	−47 26	$7\ 50.2$
7 Apr. 1940	A	−4 2	−174 37	+20 20	+126 39	+29 13	+59 33	$7\ 30.7$
19 Apr. 1958	A	0 50	−66 25	+27 40	−125 58	+30 33	+163 40	$7\ 7.2$
29 Apr. 1976	A	+7	+41	+35	−21	+32	−95	7 —
10 May 1994	A	+14	+146	+42	+81	+32	+4	7 —

TABLE 5 (*Continued*).

Date	Type[b]	Beginning Latitude	Beginning Longitude	"Noon" Latitude	"Noon" Longitude	End Latitude	End Longitude	Maximum Duration
1904 Sep. 9	T	+ 7° 53.0	−162° 49.7	− 4° 35.1	+133° 5.2	−26° 38.0	+ 69° 45.2	6^m 23.7^s
1922 Sep. 20	T	+ 5 30	− 43 17	−11 59	−106 31	−30 15	−172 36	5 58.7
1940 Oct. 1	T	+ 2 41	+ 78 30	−19 2	+ 15 36	−32 36	− 53 47	5 35.4
1958 Oct. 12	T	− 0 31	−157 26	−25 33	+139 21	−33 39	+ 66 37	5 10.7
1976 Oct. 23	T	− 4	− 31	−31	− 95	−33	−171	5 —
1994 Nov. 3	T	− 8	+ 97	−36	+ 31	−32	− 47	4 —
1905 Mar. 5	A	−52° 6.1	− 31° 13.9	−43° 19.4	−110° 0.9	−18° 19.1	−172° 33.5	7^m 58.3
1923 Mar. 17	A	−50 49	+ 76 13	−36 40	+ 3 50	−15 25	− 56 40	7 51.3
1941 Mar. 27	A	−47 54	−177 51	−29 19	+115 53	−12 43	+ 56 43	7 40.9
1959 Apr. 8	A	−43 36	− 72 25	−21 36	−133 40	−10 21	+167 52	7 25.9
1977 Apr. 18	A	−38	+ 33	−13	− 25	− 8	− 83	7 —
1995 Apr. 19	A	−31	+137	− 6	+ 82	− 6	+ 23	7 —
1905 Aug. 30	T	+50° 15.1	+ 96° 22.7	+45° 52.0	+ 12° 23.3	+18° 36.4	− 54° 49.1	3^m 45.4
1923 Sep. 10	T	+48 16	−154 18	+37 58	+128 16	+13 43	+ 63 51	3 36.9
1941 Sep. 21	T	+45 44	− 42 22	+30 18	−113 52	+ 9 58	−176 39	3 21.8
1959 Oct. 2	T	+42 35	+ 72 5	+22 57	+ 5 41	+ 7 21	− 56 16	3 1.8
1977 Oct. 12	T	+39	−171	+16	+127	+ 6	+ 65	3 —
1995 Oct. 24	T	+34	− 51	+10	−110	+ 5	−172	3 —
1906 Feb. 22	P	Antarctic						
1924 Mar. 5	P	Antarctic						
1942 Mar. 17	P	Antarctic						
1960 Mar. 27	P	Antarctic						
1978 Apr. 7	P	Antarctic						
1996 Apr. 17	P	Antarctic						

Day	Month	Year	Type	Region / Lat.	Long.	Lat.	Long.	Lat.	Long.	Duration
21	Jul.	1906	P	Antarctic						
31	Jul.	1924	P	Antarctic						
12	Aug.	1942	P	Antarctic						
		End of Saros								
19	Aug.	1906	P	Arctic						
29	Aug.	1924	P	Arctic						
10	Sep.	1942	P	Arctic						
20	Sep.	1960	P	Arctic						
2	Oct.	1978	P	Arctic						
12	Oct.	1996	P	Arctic						
13	Jan.	1907	T	+50° 25.8	− 42° 18.8	+38° 40.3	− 89° 12.1	+56° 45.2	−130° 49.3	2ᵐ 23.6
24	Jan.	1925	T	+48 18	+ 94 24	+42 9	+ 43 33	+61 28	+ 3 5	2 31.8
5	Feb.	1943	T	+47 1	−129 51	+47 11	+175 34	+66 31	+135 31	2 38.9
15	Feb.	1961	T	+46 32	+ 5 33	+53 26	− 52 50	+71 41	− 93 42	2 45.1
26	Feb.	1979	T	+47	+140	+61	+ 77	+77	+ 34	3 —
9	Mar.	1987	T	+49	− 87	+71	−154	+83	+159	3 —
10	Jul.	1907	A	−34° 32.4	+100° 30.6	−16° 57.5	+50° 25.1	−37° 21.2	+ 1° 7.0	7ᵐ 23.6
21	Jul.	1925	A	−37 32	−161 52	−25 51	+147 38	−47 29	+100 17	7 14.5
1	Aug.	1943	A	−41 48	− 61 21	−36 36	−113 43	−57 55	−158 22	6 58.6
11	Aug.	1961	A	−48 4	+ 39 2	−50 6	− 13 36	−68 48	− 51 18	6 35.3
22	Aug.	1979	A	−58	+142	−76	+ 87	−78	+ 87	6 —
1	Sep.	1997	P	Antarctic						
3	Jan.	1908	T	+10° 45.1	−154° 39.2	−11° 51.0	+145° 13.9	+ 9° 53.4	+84° 49.1	4ᵐ 12.9
14	Jan.	1926	T	+6 52	− 21 9	−10 5	− 82 45	+14 28	−141 56	4 10.7
25	Jan.	1944	T	+3 23	+111 59	− 7 23	+ 49 15	+18 48	− 9 23	4 8.9
5	Feb.	1962	T	+0 35	−115 51	− 3 44	−179 14	+22 53	+121 58	4 8.0
16	Feb.	1980	T	−1	+ 15	+ 1	− 48	+27	−108	4 —
26	Feb.	1998	T	−2	+144	+ 6	+ 81	+30	+ 19	4 —

TABLE 5 (*Continued*).

Date	Type[b]	Beginning Latitude	Beginning Longitude	"Noon" Latitude	"Noon" Longitude	End Latitude	End Longitude	Maximum Duration
June 28 1908	A	+4° 39.3	+129° 56.8	+31° 27.3	+66° 55.3	+10° 1.0	+1° 8.0	4^m 1^{s}1
Jul. 10 1926	A	+4 12	−132 4	+25 36	+165 6	+1 27	−103 29	3 51.1
Jul. 20 1944	A	+3 30	−33 25	+19 0	−95 46	−6 57	−154 20	3 42.0
Jul. 31 1962	A	+2 32	+67 1	+11 49	+5 20	−14 57	−51 30	3 32.8
Aug. 10 1980	A	+1	+169	+4	+108	−23	+52	3 —
Aug. 22 1998	A	−1	−87	−4	−147	−29	+155	3 —
Dec. 23 1908	C	−22° 46.1	+73° 31.7	−53° 46.0	−2° 27.6	−31° 54.4	−86° 1.9	0^m 11^{s}3
Jan. 3 1927	A	−27 1	−156 31	−52 49	+124 34	−27 33	+45 13	0 46.2
Jan. 14 1945	A	−31 21	−26 39	−51 18	−107 55	−23 37	+176 45	0 53.0
Jan. 25 1963	A	−35 23	+101 47	−48 59	+19 17	−19 58	−52 19	1 1.3
Feb. 4 1981	A	−39	−132	−45	+146	−16	+78	1 —
Feb. 16 1999	A	−41	−8	−41	−88	−13	−154	1 —
June 17 1909	C	+50° 1.7	−81° 31.6	+88° 22.1	+172° 39.1	+60° 23.2	+41° 58.7	0^m 23^{s}2
June 29 1927	T	+46 29	+16 14	+78 25	−83 55	+51 1	+168 34	0 50.2
Jul. 9 1945	T	+44 23	+115 57	+70 3	+20 2	+41 43	−72 33	1 15.6
Jul. 20 1963	T	+43 6	−142 12	+62 17	+125 44	+33 2	+43 42	1 39.7
Jul. 31 1981	T	+42	−40	+54	−127	+25	+159	1 —
Aug. 11 1999	T	+41	−65	+46	−18	+17	−87	2 —
Dec. 12 1909	P	Antarctic						
Dec. 24 1927	P	Antarctic						
Jan. 3 1946	P	Antarctic						
Jan. 14 1964	P	Antarctic						
Jan. 25 1982	P	Antarctic						
Feb. 5 2000	P	Antarctic						

Day	Mo.	Year	Type							
8	May	1910	T	−72° 37′.1	−112° 1′.5	−47° 26′.4	+126° 30′.0	−46° 31′.2	−156° 12′.6	4ᵐ 14ˢ.0
19	May	1928	T	−67 11	−12 18	−63 17	−22 25	−58 24	−29 14	4 —
30	May	1946	P	Antarctic						
10	June	1964	P	Antarctic						
21	June	1982	P	Antarctic						
1	Jul.	2000	P	Antarctic						
1	Nov.	1910	P	Arctic						
12	Nov.	1928	P	Arctic						
23	Nov.	1946	P	Arctic						
4	Dec.	1964	P	Arctic						
15	Dec.	1982	P	Arctic						
25	Dec.	2000	P	Arctic						
28	Apr.	1911	T	−36° 47′.9	−148° 37′.3	−0° 36′.3	+154° 43′.8	+11° 5′.9	+90° 2′.3	4ᵐ 57ˢ.2
9	May	1929	T	−36 46	−34 57	−0 54	−89 35	+4 48	−153 3	5 7.3
20	May	1947	T	−36 30	+77 46	−1 58	+24 40	−2 12	−36 58	5 14.1
30	May	1965	T	−36 18	−170 51	−4 24	+137 10	−9 55	+77 49	5 15.9
11	June	1983	T	−36	−60	−7	−111	−18	−168	5 —
21	Oct.	1911	A	+44° 44′.8	−60° 31′.1	+10° 34′.5	−117° 32′.7	−7° 50′.2	−117° 29′.0	3ᵐ 50ˢ.6
1	Nov.	1929	A	+43 27	+54 42	+8 23	+0 43	−3 45	−59 10	3 57.7
12	Nov.	1947	A	+41 6	+172 59	+6 6	+121 5	+0 34	+61 39	4 2.9
23	Nov.	1965	A	+37 53	−65 30	+3 56	−116 24	+5 1	−175 4	4 4.9
4	Dec.	1983	A	+34	+58	+2	+8	+10	−50	4 —
17	Apr.	1912	C	+5° 3′.5	+61° 11′.4	+46° 52′.9	0° 57′.6	+57° 19′.8	−89° 48′.4	0ᵐ 1ˢ.6
28	Apr.	1930	C	+3 32	+172 57	+45 41	+112 22	+50 46	+22 44	0 1.5
9	May	1948	A	+2 33	−77 8	+43 57	−138 8	+43 40	+135 10	0 51.7
20	May	1966	A	+1 52	+30 14	+41 33	−31 10	+36 1	−113 35	0 56.2
30	May	1984	A	+1	+136	+38	+74	+28	−3	1 —
10	Oct.	1912	T	+3° 43′.1	+92° 28′.6	−34° 56′.9	+33° 11′.2	−52° 28′.1	−47° 14′.5	1ᵐ 54ˢ.9
22	Oct.	1930	T	+4 17	−145 48	−36 6	+154 45	−48 6	+72 6	1 55.3
1	Nov.	1948	T	+3 42	−22 3	−37 21	−81 58	−43 23	−165 27	1 55.8
12	Nov.	1966	T	+2 00	+104 9	−38 39	+43 14	−38 29	−39 53	1 57.3
22	Nov.	1984	T	+0	−128	−39	+170	−33	+88	2 —

TABLE 5 (*Continued*).

Date	Type[b]	Beginning Latitude	Beginning Longitude	"Noon" Latitude	"Noon" Longitude	End Latitude	End Longitude	Maximum Duration
6 Apr. 1913	P	Arctic						
18 Apr. 1931	P	Arctic						
28 Apr. 1949	P	Arctic						
9 May 1967	P	Arctic						
19 May 1985	P	Arctic						
31 Aug. 1913	P	Arctic						
12 Sep. 1931	P	Arctic						
Last Eclipse of Saros								
29 Sep. 1913	P	Antarctic						
11 Oct. 1931	P	Antarctic						
21 Oct. 1949	P	Antarctic						
2 Nov. 1967	T	$-56° \ 22'$	$+18° \ 46'$			$-67° \ 18'$	$+39° \ 58'$	—
12 Nov. 1985	T	-52	$+146$			-70	$+164$	—
24 Feb. 1914	A	$-77° \ 35\!.\!0$	$+30° \ 54\!.\!6$	$-62°$	$+115°$	$-42° \ 51\!.\!2$	$+90° \ 41\!.\!5$	—
7 Mar. 1932	A	$-74 \ 37$	$+179 \ 13$	-60	-140	$-47 \ 8$	$-152 \ 29$	$5^m \ 19^s\!.\!1$
18 Mar. 1950	A	$-72 \ 14$	$-48 \ 0$	$+61 \ 0$	$-40 \ 54$	$-49 \ 35$	$-34 \ 30$	—
28 Mar. 1968	P	Antarctic						
9 Apr. 1986	P	Antarctic						
21 Aug. 1914	T	$+71° \ 23\!.\!4$	$+121° \ 11\!.\!9$	$+70° \ 52\!.\!5$	$-2° \ 4\!.\!2$	$+23° \ 44\!.\!7$	$-70° \ 35\!.\!8$	$2^m \ 14^s\!.\!5$
31 Aug. 1932	T	$+79 \ 36$	$-109 \ 16$	$+78 \ 36$	$+109 \ 10$	$+28 \ 27$	$+40 \ 59$	1 44.8
12 Sep. 1950	T	$+85 \ 10$	$+66 \ 42$			$+34 \ 23$	$+154 \ 35$	1 13.7
22 Sep. 1968	T	$+80$	-108			$+42$	-90	1 —
3 Oct. 1986	T	$+66$	$+26$			$+56$	$+28$	1 —

Day	Month	Year	Type												Duration
13	Feb.	1915	A	−35° 49.7	− 42° 35.1	−26° 29.7	−117° 57.1	+13° 9.8	−174° 53.8	2m 17s.1
24	Feb.	1933	A	−39 25	+ 79 9	−23 56	+ 5 5	+14 28	− 52 18	1 55.0
7	Mar.	1951	A	−42 32	−161 18	−21 28	+126 52	+14 35	− 68 40	1 38.0
18	Mar.	1969	A	−45	− 44	−19	−112	+13	−172	1 —
29	Mar.	1987	C	−47	+ 71	−17	+ 6	+11	− 54	1 —
10	Aug.	1915	A	+23° 12.3	−129° 39.2	+16° 32.4	+161° 35.0	+21° 55.9	+106° 31.3	1m 57s.9
21	Aug.	1933	A	+30 11	− 24 38	+17 55	− 94 48	−20 31	−150 38	2 18.0
1	Sep.	1951	A	+36 19	+ 80 56	+18 42	+ 10 28	−18 38	− 46 9	2 43.5
11	Sep.	1969	A	+41	−173	+19	+117	−16	+ 60	2 —
23	Sep.	1987	A	+46	− 68	+19	−135	−13	+167	2 —
3	Feb.	1916	T	+ 7° 20.8	+121° 11.9	+15° 57.2	+ 61° 56.5	+49° 23.8	+ 9° 50.2	2m 36s.2
14	Feb.	1934	T	+ 3 55	−107 50	+19 22	−168 2	+52 26	+136 41	2 52.7
25	Feb.	1952	T	+ 0 46	+ 21 14	+22 38	− 39 15	+54 24	− 99 12	3 9.5
7	Mar.	1970	T	− 2	+149	+25	+ 88	+55	+ 23	3 —
18	Mar.	1988	T	− 4	− 86	+28	−146	+54	+143	3 —
29	Jul.	1916	A	−28° 44.6	− 89° 32.1	−36° 53.7	−141° 41.5	−63° 35.6	−178° 36.5	6m 24s.9
10	Aug.	1934	A	−19 36	− 10 47	−33 9	− 43 12	−62 31	− 87 53	6 34.3
20	Aug.	1952	A	−11 37	+111 23	−30 35	+ 56 12	−61 9	− 4 12	6 41.1
31	Aug.	1970	A	− 5	−147	−29	+157	−59	+ 98	7 —
11	Sep.	1988	A	+ 1	− 44	−28	−101	−56	−165	7 —
24	Dec.	1916	P	Antarctic						
5	Jan.	1935	P	Antarctic						
End of Saros										
22	Jan.	1917	P	Arctic						
3	Feb.	1935	P	Arctic						
14	Feb.	1953	P	Arctic						
25	Feb.	1971	P	Arctic						
7	Mar.	1989	P	Arctic						

TABLE 5 (*Continued*).

Date	Type[b]	Beginning Latitude	Beginning Longitude	"Noon" Latitude	"Noon" Longitude	End Latitude	End Longitude	Maximum Duration
19 June 1917	P	Arctic						
30 June 1935	P	Arctic						
11 Jul. 1953	P	Arctic						
22 Jul. 1971	P	Arctic						
End of Saros								
18 Jul. 1917	P	Antarctic						
30 Jul. 1935	P	Antarctic						
9 Aug. 1953	P	Antarctic						
20 Aug. 1971	P	Antarctic						
31 Aug. 1989	P	Antarctic						
13 Dec. 1917	A	$-59°$ 1$'$.9	$+87°$ 52$'$.7	$-89°$ 56$'$.6	$+142°$ 12$'$.8	$-56°$ 7$'$.8	$-155°$ 41$'$.2	—
25 Dec. 1935	A	-62 18	-134 59	-87 43	-93 14	-53 14	-25 6	—
5 Jan. 1954	A	-65 32	$+3$ 21	-85 2	$+31$ 15	-50 45	$+105$ 49	—
16 Jan. 1972	A	-69	$+143$	-81	$+156$	-49	-123	—
26 Jan. 1990	A	-71	-74			-48	$+7$	—
New Saros								
17 June 1928	P	Arctic						
28 June 1946	P	Arctic						
9 Jul. 1964	P	Arctic						
20 Jul. 1982	P	Arctic						
31 Jul. 2000	P	Arctic						

[a] Data from May 28, 1900 through September 20, 1960 taken from *American Ephemeris and Nautical Almanac.* Data for February 15, 1961 through November 2, 1967 from U. S. Naval Observatory Circulars. Data for 1967–2000, from Oppolzer, *Canon der Finsternisse.* The data from Oppolzer are, in general, less accurate than those from the *Almanac.* The maximum durations, for 1967–2000, are estimated by extrapolation.

[b] P = Partial, T = Total, A = Annular, C = Central (i.e., annular, then total, then annular).

can calculate the time of eclipse to within a second or two and fix the location of the shadow track with an accuracy better than 1000 feet. Small uncertainties arise, however, from vagaries in the moon's motion and in the rotation of the earth itself. In fact, data of observed eclipse times have assisted astronomers to more precise study of the moon's orbital motion.

The maximum number of eclipses in a year is seven, including five of the sun and two of the moon or four of the sun and three of the moon. The minimum number is two, both of the sun. The total number of hours of total eclipse during the entire century is only 4 hours 50 minutes, an average of 2.9 minutes per year. If an astronomer, with fifty years of useful observing life, went to every eclipse, stationed himself in the path of maximum duration, and had clear weather for all eclipses, his total observing time would be only 2½ hours. These figures indicate the importance of such observations and also emphasize how the coronagraph has multiplied the eclipse life of an astronomer by many thousandfold.

$\odot$

Eclipses of Historical Interest

The interested amateur, following the rules previously given, will find it a relatively simple matter to predict with some degree of accuracy when and (occasionally) where eclipses will occur or have occurred in the past. The 29-year period, applied to the previously mentioned 1905 eclipse, predicts the eclipses of 10 August 1934, 20 July 1963, 30 June 1992, and 10 June 2021. From this last date, five applications of the 521-year cycle carry us back to 10 June 585 B.C., or rather to 28 May, since we must intercalate the 13 days difference now existing (14 days after A.D. 2100) between the Gregorian and Julian calendars.

This ancient eclipse in 585 B.C., thus simply calculated, was one of the most famous in all history. The Greek scientist, Thales of Miletus, may have forecast its occurrence. The Medes and the Lydians, waging a drawn-out, five-year war, were so impressed by the eclipse that they threw down their arms and made a lasting peace, firmly cemented by the bond of a double marriage.

A second notable eclipse, this time of the moon, occurred in 413 B.C. The battle of Syracuse was in progress. The Athenian fleet, bottled up in the harbor at Syracuse, had made plans for a night

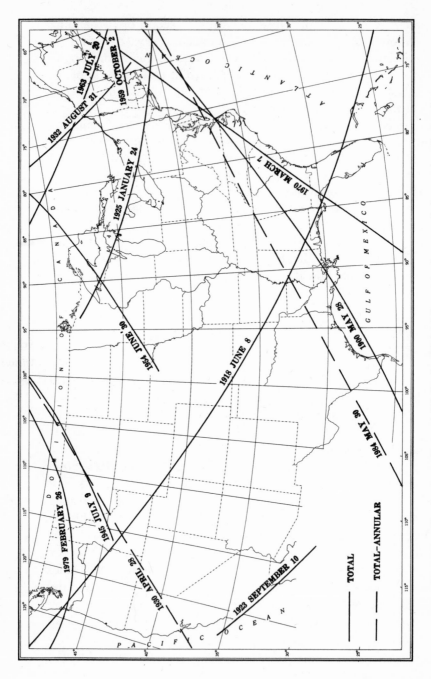

Fig. 156. Total solar eclipses visible in the United States in the twentieth century. Paths of eclipses after 1970 are approximate. (U. S. Naval Observatory.)

escape. The eclipse of the moon, which happened on the fateful night, so frightened the men that they decided to postpone the scheduled flight. The eclipse, so the superstitious Athenians believed, was an ill omen. And so it turned out. The delay enabled the Syracusans to strengthen their own forces. In the ensuing battle the Athenians lost their entire fleet and their army. More than once has an eclipse altered the course of history.

The earliest eclipse of which we have written record probably occurred on 22 October 2137 B.C. The ancient Chinese classic, *Shu Ching,* states that the two royal astronomers, Hsi and Ho, were so "drunk in excess of wine" that they failed to warn the populace of the occurrence. The eclipse, thus unexpected, so frightened the people that they stampeded through the streets, beating drums and shooting fireworks to frighten away the dragon devouring the sun. Now, according to ancient Chinese law concerning the duties of astronomers in the matter of eclipse prediction, any error was of the gravest consequence. If the astronomer's forecast were "behind the time he would be hanged without respite or ahead of the time he would be killed without reprieve." We are not certain that Ho and Hsi were summarily executed; the records are obscure concerning their fate.

The superstitious awe of eclipses has not completely passed away. As recently as 1914, when a total eclipse occurred in the Ukraine, the Russian peasants thronged to the churches in terror. At the eclipse of 1922, the Australian aborigines got the idea that the astronomers were attempting to throw a net over the sun and capture it. And many persons, otherwise well informed, accept the astrological belief that an eclipse is an ill omen.

⊙

Scientific Results

Ancient eclipses have served a useful scientific purpose. For, by comparing the predicted path of totality with the observed, we may test for possible variations in the earth's speed of rotation. Astronomers have thus found that the day is gradually lengthening, by the tiny amount of 0.001 second in a century. The cause of the slowing down is attributed to the frictional effect of ocean tides, particularly in the shallower seas.

On numerous occasions, in earlier chapters, I have referred to solar studies made at times of total eclipse. The spectra of the corona, of the chromosphere, and of the extreme solar limb are best obtained on such occasions.

For flash-spectrum recording, Mitchell, veteran of many eclipses, has employed a simple concave grating pointed directly at the sun. This instrument gives a series of crescents like those shown in Fig. 56. Mitchell, watching the progress of the eclipse, selects the precise moment for opening and closing the camera shutters—a feat requiring judgment and careful timing.

W. W. Campbell, another of the regular eclipse followers, employed procedures similar to those of Mitchell. In addition, he devised the so-called moving-plate recording. Campbell set, parallel to the spectrum, a narrow slit that admitted a small portion of light from the center of each crescent. Then, as the eclipse progressed, a continuously moving plate or film recorded the changing character of the spectrum. A portion of Campbell's beautiful 1905 flash spectrum appears in Fig. 160, with the normal solar spectrum added for comparison.

Campbell's moving-plate method gives superior results when the portion of the crescent falling on the slit has no prominences super-

Fig. 157.   Some eclipse equipment from the Harvard–Massachusetts Institute of Technology expedition to the U.S.S.R., 19 June 1936. W. R. Brode, left center.

Fig. 158. Eclipse equipment, Fryeburg, Maine, 31 August 1932. The instruments are (left to right): flash spectrum, jumping-film cameras; flash spectrum, moving-plate cameras; coronal spectrographs; coronal cameras. (Lick Observatory.)

posed on the chromosphere. When prominences occur, however, the lines are often doubled or blurred. For that reason, I prefer a refinement of Mitchell's method, where a series of exposures, automatically spaced and controlled, reproduce the complete history of the eclipse spectra.

The invention of the coronagraph has somewhat altered the relation of eclipses to scientific research. Certainly, the general programs of eclipse expeditions should undergo something of a change. Scientists should leave to the coronagraph all studies well suited to that instrument and concentrate their attention on problems requiring a sky especially dark.

Chief among the latter, which include several problems yet unsolved, are: the distribution of light in the continuous spectrum of the corona, especially in the faint outer regions; the structure and polarization of faint coronal streamers; the relation of the external to the internal corona; the recording of faint lines of the chromospheric flash spectrum with high-dispersion spectrographs; study of the flash spectrum in the far ultraviolet and infrared; miscellaneous technical studies concerned with faint details that are ordinarily suppressed by scattered sky illumination in the coronagraph.

Eclipses provide an opportunity for locating the precise areas of the solar surface whose ultraviolet radiation is responsible for the earth's ionosphere. Thus radio studies of the ionosphere itself, made during eclipses, should show and actually do show relations between the sun and the earth's upper atmosphere. For such researches,

Fig. 159.   William Wallace Campbell. (From a portrait by Peter Van Valken-burgh, June 1930.)

annular and partial eclipses are also useful, whereas such eclipses are of limited value in ordinary photographic studies. I shall detail some of these problems in the last chapter.

⊙

One of the most important problems still requiring further study at *Tests of the* eclipses is the deflection of starlight in the sun's gravitational field. *Einstein* Einstein, developing his famous theory of relativity, predicted as *Theory* one of its consequences that a ray of starlight grazing the sun would be turned through a small angle. Hence the image of a star should be displaced radially outward by an amount ranging downward from 1.74 seconds of arc, the value at the sun's edge.

The angle is extremely minute; 1.74 seconds corresponds to the angle subtended by a dime at a distance of about a mile and a half. Even so, astronomers are accustomed to measuring quantities very much smaller. Here, however, one must take special precautions. The scientist sets up his equipment—a powerful photographic telescope—within the belt of totality. He employs heavy photographic plates with special emulsions designed for stability and freedom from shrinkage. Only a few eclipses are satisfactory for the investigation—those where the sun is surrounded by a number of

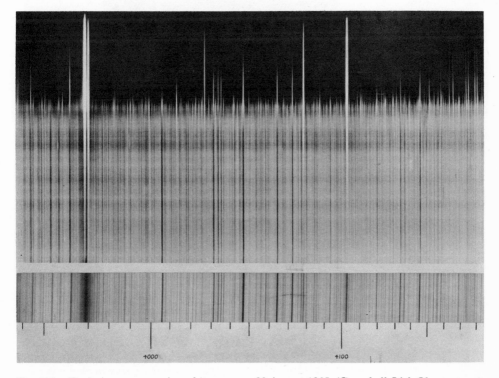

Fig. 160.   Flash spectrum, moving-plate camera, 30 August 1905. (Campbell, Lick Observatory.)

sufficiently bright stars. After taking the eclipse plates, the scientist must wait several months, until the sun has moved out of that portion of the sky. Then, at night, he rephotographs the same region and compares the positions of the stars on the two plates.

The measurement of the plates occupies months—sometimes years—of study. Corrections are usually necessary because of the interval between the eclipse and the comparison photographs, and this further complicates the problem. But finally one determines the deflections. So difficult is the program that there have been very few successful expeditions, and the results are somewhat dis-

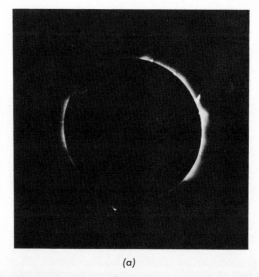

(a)

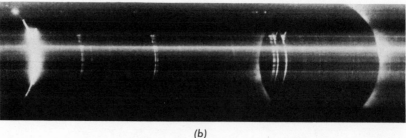

(b)

Fig. 161.   Solar eclipse, 25 February 1952: (a) inner corona and bright prominence spike (U. S. Naval Research Laboratory); (b) spectrum, showing bright prominence spike, especially in the three magnesium lines. The circle is due to the green coronal line, with Hβ to far left (High Altitude Observatory).

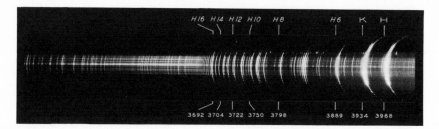

Fig. 162.   Ultraviolet flash spectrum, 25 February 1952. (High Altitude Observatory.)

cordant. The records that Campbell and Trumpler obtained in Australia (1922) agree closely with Einstein's predicted value. Van Biesbroeck found values of 2.01 and 1.70 seconds at the eclipses in 1947 and 1952, respectively, but he cautions that the latter figure is uncertain because high winds shook his instrument during the eclipse. Freundlich in 1929 obtained a value in excess of 2 seconds. From the 1936 eclipse Michailov found a value as large as 2.70 sec. The present mean from all the measures appreciably exceeds Einstein's predicted value. In view of these discordances further observations are clearly necessary.

The three astronomical tests of the Einstein theory—the deflection of light in a gravitational field, minute shifts in wavelength of the lines of the solar spectrum (measured by St. John), and the slow

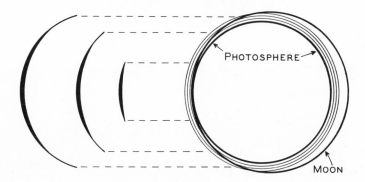

Fig. 163.   Geometry of formation of the flash spectrum. When the moon is just tangent to one limb of the photosphere, the latter's continuous spectrum cannot be seen, so the bright emission lines of the chromosphere flash out at various wavelengths. The crescent lengths are greater for higher chromospheric levels. The relative scale is not correct. (*Sky and Telescope.*)

advance of the perihelion of the orbit of the planet Mercury—appear to be in fairly good accord with the predictions, and one may state that experiment seems to confirm the Einstein theory.

⊙

**Vulcan, the
Intra-Mercurial
Planet**

Another type of study attempted at eclipse expeditions—though now seldom tried—has been the search for a planet wthin the orbit of Mercury. A planet so placed would be so close to the sun that we could scarcely expect to observe it in the brilliant twilight skies. One would expect to discover this hypothetical planet only by seeing it as a black dot against the solar disk or by viewing it as a star at the time of totality.

Fig. 164.   Einstein camera, Wallal, Australia, 21 September 1922. (Lick Observatory.)

Fig. 165.   Einstein plate, 1922 eclipse. The circles indicate the positions of stars measured for Einstein deflection. (Lick Observatory.)

The idea of a possible intra-Mercurial planet first arose in 1859, when a French amateur astronomer, Dr. Lescarbault, a physician living at Orgères, noted a black spot moving rapidly across the solar disk. No one has seen it since, though many have tried. The object is generally called Vulcan. The many attempts to observe it have all failed. Most astronomers now agree (a) that the object seen by Lescarbault was not an intra-Mercurial planet, and (b) that, if such a planet exists, it must be of extremely small size.

⊙

Eclipses and Animal Life   Another interesting aspect of eclipses is their effect upon animals and insects. There have been many scattered reports concerning the reactions of animal life, but the eclipse of 1932 was the first where scientists made a definite coordinated study of the behavior of animals during the phenomenon. The previous fragmentary records had indicated that birds and animals appeared to be frightened by the oncoming shadow. There was no mention of species involved or checks to determine their normal behavior. If, for example, a blue heron happens to take wing during totality, what is there to prove that he would not have flown anyway?

The collected records of the 1932 eclipse showed that most of the birds and animals apparently interpreted the darkening either as the coming of night or as the onrush of a storm. The roosting of chickens at eclipse time has been accepted as a fact, almost from antiquity. Certainly chickens go into the coop, but no one, until 1932, thought of making a further study. A check-up at this eclipse revealed that most of the birds were huddled together on the floor of the enclosure, few having taken to the roosts. Many never even entered the coop.

Fig. 166.   George Van Biesbroeck, of Yerkes Observatory, with an Einstein camera, 20 May 1947. (National Geographic Society.)

One true anecdote refers to the eclipse of 28 May 1900, whose path of totality cut through Georgia. There the Negro postman, who delivered mail to the astronomers, was impressed most of all by the forecast that the chickens would enter the coop. Totality found him in the chicken yard. Afterward hastening to report, "Boss Campbell," he said to the director of the expedition from Lick Observatory, "dem chickens done roost jus' like you say." And then, as an afterthought, he asked, "How long you know dem chickens gwine roost?" Dr. Campbell replied, "Oh, I've known that for two or three years at least." The postman stared incredulously, and then laughed. "Go 'long, Boss," he said. "You can't fool dis here pusson. Dem chickens wusn't bo'n till las' spring!"

Observations showed that many domestic animals pay no attention whatever to eclipse phenomena. The unnatural expectancy and excitement of their masters probably account completely for

Fig. 167.   Battery of cameras used in search for intra-Mercurial planet, Vulcan. (Lick Observatory.)

the sense of uneasiness displayed by dogs. The only authenticated case of apparent fright was that of a cat, which climbed a clothes pole and howled throughout totality, the while looking directly at the eclipsed sun.

Insects showed more reaction than almost any other type of life. Flying ants swarmed from the ground on a nuptial flight, an action traceable to the lowered temperature rather than to the dimness of the solar light. Bees returned to the hive and rarely ventured out again, even after the eclipse was over. In one case, bees, in a sudden flurry of activity, drove the drones from the hive, a customary procedure in late fall but unprecedented as early as 31 August. The observer of the phenomenon ventures, with due reserves, the theory that the sudden advent of darkness at 3:15 p.m. may have fooled the bees into concluding that short winter days were at hand, and driven them to act accordingly.

$$\odot$$

Eclipse
Expeditions   The preparation of an eclipse expedition entails a vast amount of work. There are the problems of exact calculations of the path, the examination of weather records, the picking of a site, questions of transportation, and the securing of necessary facilities such as electricity or water. Then there are the devising of a program and the building of appropriate instruments. The more inaccessible the site —and eclipses seem to have a perverse habit of hitting desolate regions—the greater is the necessity of providing for all possible contingencies.

The planning of a real eclipse expedition occupies months of work. There are, of course, many scientists whose desire simply to see the event leads them to travel with minimum equipment, such as a small camera to photograph the corona. But an expedition, fully equipped to study radio, spectrographic, and similar problems, may carry from 10 to 20 tons of material. All of this equipment must be assembled, often under adverse climatic conditions, with an absolute and immovable deadline: the eclipse itself.

The strain of the work, the inevitable delays, and the worry about weather conditions complicate the problem. The moment of totality often finds the astronomer so busy that he is fortunate to get even two free seconds for a glimpse of the corona. Hours or days of developing and standardizing the photographic records follow. The laborious packing and trip home are an anticlimax. And if the weather at the time of the eclipse is cloudy, the months of effort are completely lost. Only the radio portion of the research program can be salvaged, because radio waves can penetrate the overcast.

Although one may employ an airplane for certain types of observations, the great size and weight of the more important and delicate equipment precludes its use from the air. Such devices must be pointed with great precision and they are very sensitive to vibration. Nevertheless, a ready plane is good insurance against a complete failure through blackout by clouds, even though the results so obtained are usually far less valuable than those taken from ground-based equipment.

11

Atomic Energy and the Solar Interior

Our next task is one of astrophysical prospecting, to explore what lies below the sun's shining surface. The problem is not as impossible as might appear at first sight. We already have several clues. As recounted in Chapter 1, we can determine the mass of the sun from gravitational effects. We also know the quantity and the quality of its radiation. We have measured the radius of the sun and we know the characteristics of the solar surface, including the rate of rotation. Our tools for prospecting are the laws of mathematics and physics, which we assume to be applicable everywhere in the universe. Let us see what we can deduce from the data at hand.

257

The total heat output of the sun is about 3×10^{33} calories per year. Its mass is 2×10^{33} grams. Thus, the average annual output is 3 calories for every 2 grams or 1.5 calories per gram. This figure may not sound very large. But when we consider that the sun has been in existence for more than a billion (10^9) years, and that its output and mass have probably not varied appreciably over that interval, the total output of more than 10^9 calories per gram seems enormous. One gram of the purest coal and oxygen, mixed in correct proportion for complete combustion, will release only 2200 calories. Thus if the sun consisted of oxygen and carbon, it could exist only 1500 years or so.

Astronomers long ago recognized the limitation of chemical heat and looked for other sources of energy. Some suggested that meteors might help. But, in the neighborhood of the earth, these shooting stars contribute only a negligible fraction of the total energy we receive. They may be considerably more numerous close to the sun, but calculations show the meteoric source to be utterly inadequate.

For years, most scientists accepted Helmholtz's proposal that slow compression was responsible for keeping the sun going. Compression automatically heats a gas, whereas expansion cools it. Suppose that the sun was once at least as large as the solar system. As gravitation pulled the atoms together, the material would gradually heat itself. Helmholtz showed that the maximum age of the sun, on this hypothesis, would not be greater than 50 million years. In fact he firmly forbade the geologists to find evidence for a greater age of the earth. But eventually the latter, in noncooperative spirit, proceeded to dig up indisputable evidence for a still greater age, at least 2000 million (2×10^9) years. The best and most recent figure is 4500 million years, based on a determination of the age of the oldest terrestrial rocks.

Faced with this difficulty, astronomers were forced to accept what seemed to be the only remaining hypothesis: that matter was convertible into energy. This idea came into being long before Einstein or the atomic bomb. Sir Isaac Newton had once speculated on the possibility of changing "bodies into light or light into bodies."

Einstein, however, provided the quantitative key to the process in his now famous equation,

$$E = Mc^2,$$

where E is the energy in ergs equivalent to a mass M in grams, and c is the velocity of light, 3×10^{10} cm/sec. (An erg is the physicist's unit of energy. The unit is small. A pencil, falling from the desk to the floor, will acquire several million ergs—almost a calorie—of energy.) Rewriting the equation, and expressing E in calories, we get

$$E = 2.15 \times 10^{13} \ M \text{ calories.}$$

One gram—half a thimbleful—of water, if completely converted into energy, would furnish 2×10^{13} calories, as much as 20,000 tons of coal. Here, indeed, would be an ample source of power, if the sun could turn it on or control it. It could keep the sun going for 10 million million (10^{13}) years. Of course, we may not be able to annihilate matter completely. But if we could convert only a fraction into energy we should be far ahead of chemical or combustion processes.

Actually two ways have been discovered for getting this energy out of matter: by tearing down or splitting heavy atoms, and by assembling lighter atoms into heavier components. We have known something about the first of these processes since 1895, when Becquerel found that a compound of uranium gave off rays that would pass through paper or thin metal sheets. Uranium itself has been known since 1789.

This new property, called radioactivity, fired the imaginations of Pierre and Marie Curie. Madame Curie found that another known element, thorium, possessed similar properties. She also discovered polonium (named for her native Poland), in July 1898. She and her husband isolated radium five months later. These atoms, all heavy ones near the end of the periodic table, are naturally unstable.

A radioactive atom may emit alpha particles (helium nuclei) or beta particles (electrons). The former are positively charged; the latter, negatively. The atom may also send out a gamma ray, which is similar to an x-ray, that is, light of extremely short wavelength. One cannot speed up or slow down the process of natural radioactive disintegration.

☉

Structure of
Atomic Nuclei

In Chapter 3 we talked a good deal about atoms. But there our chief concern was with their exteriors—the superstructure made of negative electrons. The nucleus, with its positive charge and most of the mass, appeared only as a core that held the atom together. Now we shall see how nuclei themselves are constructed.

Small as the atom is—about 10^{-8} cm in diameter—we must visualize the nuclear core as being far smaller: 10^{-12} cm or so across. If the small ash tray on my desk were the nucleus, the electron would be a schoolboy's marble flying through space about a mile away. The atom is mostly empty space. If I could pick the atoms apart with submicroscopic tweezers—as a biologist dissects a worm—and pile up the fragments in a heap, the seemingly closely packed material from the entire Empire State Building would just about fill the space of an ordinary-sized pinhead! Only the mysterious forces of electric attraction and repulsion keep the universe from collapse.

There are certain building blocks available for constructing the nucleus; they are listed in Table 6.

TABLE 6.   NUCLEAR BUILDING BLOCKS.

Name	Charge	Mass (grams)
Electron	Negative	9.1083×10^{-28}
Positron	Positive	9.1083×10^{-28}
Proton	Positive	1.6724×10^{-24}
Antiproton	Negative	1.6724×10^{-24}
Neutron	Zero	1.6747×10^{-24}
Deuteron	Positive	3.3429×10^{-24}
Alpha particle	Positive	6.6430×10^{-24}

In addition to these particles, we also have the mysterious charged mesons, heavier than electrons and lighter than protons. Short-lived phenomena, the mesons appear to be bundles of energy. Although mesons most commonly result from cosmic rays, they have also been produced in the laboratory.

I call your attention to the curious asymmetry of the above table. No philosopher would have deduced it from metaphysical principles. Although the electron and positron appear to be opposite counterparts, the latter is inherently unstable. Its lifetime is uncommonly short. In its rush through space it dashes into the nucleus of the first atom it encounters. There it unites with a neutron, to form a proton.

Thus, we have a reaction of the sort

$$\text{neutron} + \text{positron} = \text{proton},$$

and its inverse:

$$\text{proton} - \text{positron} = \text{neutron}.$$

This last equation implies that protons are the primary source of positrons. There is evidence also for the existence of the reactions

$$\text{proton} + \text{electron} = \text{neutron}$$

and

$$\text{neutron} - \text{electron} = \text{proton}.$$

And if a positron and an electron collide they annihilate one another, with the emission of a pair of gamma rays to carry off the released energy and momentum.

Scientists have recently found evidence for a negative proton, the antiproton. Such a particle could capture an orbitary positron to form an oppositely charged counterpart of the hydrogen atom. Indeed, there is no apparent reason why heavier atoms, composed entirely of antimatter, should not exist—somewhere. But a universe so constructed would be violently unstable if it chanced to encounter one made of ordinary matter.

From Table 6 one might conclude that a heavy nucleus contained many varieties of basic construction units. We find, however, that there are only two—the proton and the neutron. The deuterons and alpha particles are themselves atomic nuclei composed of these units.

The deuteron consists of one proton and one neutron. It possesses, therefore, a single positive charge, like the proton. In consequence, it can hold but a single electron in one of the possible exterior positions. When this deuteron nucleus captures an outer electron, the resulting neutral atom looks very much like hydrogen, which also has a single exterior electron. For chemical purposes, we deal mainly with the outer electronic shell, so that both of these atoms are hydrogen in the chemical sense. The double-weight atom, which we call deuterium or heavy hydrogen, is an *isotope* of the normal hydrogen. An isotope is an atom that is chemically equivalent to another, but whose nuclear mass is different.

We can produce also an isotope of hydrogen with mass three. This nucleus, known as tritium, contains one proton and two neutrons. And one might think we could go on indefinitely adding any number of neutrons, getting hydrogen isotopes of masses 4, 5, 6, and so on. But we cannot, and even isotope 3 is somewhat unstable, tending, after an average life of about 12 years, to emit an electron and decay into the rare isotope 3 of helium. The common variety of helium nucleus, isotope 4, is identical with the alpha particle, and contains two protons and two neutrons. Since protons and neutrons have nearly the same mass, a helium nucleus weighs about four times as much as a proton. Helium 5 is so unstable that we can scarcely say it really exists. For the lighter atoms of the chemical series, the most stable forms are those whose numbers of protons and neutrons are approximately equal. In the heavier atoms, the number of neutrons in the nucleus tends to exceed that of protons.

As yet, we know little of the forces that hold the nucleus together. The mysterious mesons, which probably play a significant role, are sometimes popularly referred to as "atomic glue." Protons have positive charges and hence should repel one another violently. But proximity in the nucleus, aided by the presence of neutrons and mesons, mysteriously turns the repulsive into a cohesive force. We may imagine that the nucleus has a sort of rim, like my ash tray, halfway out to its edge. A marble inside this rim tends to stay inside. A marble outside tends to roll down the sloping exterior.

Let us pursue the ash-tray analogy a little further. Suppose that I try to build up heavier nuclei by rolling marbles up the sloping sides of the tray into the interior. If I roll the marble too slowly, it climbs the side and then stops and rolls back. If I roll it too hard it bounds over and out on the other side. But if I give it just the right energy, the marble may remain inside.

Friction makes this experiment a little easier to perform than in the actual atomic case. Without the aid of friction, I could never make a marble stay in an empty ash tray. If it happens to strike another marble already in the tray, the condition is different. For it then divides its energy and, even in the absence of friction, the marbles might remain inside.

In one respect the analogy fails completely. The sloping edge of the atomic nuclear wall is not impenetrable. A particle may roll very near to the top, but with insufficient energy to jump over.

And then, presto! As if by magic, it tunnels through the wall into the nucleus.

When a nucleus has captured a fast particle, the excited neutrons and protons, bouncing around inside, have several ways of losing this excess energy of motion. The nucleus may eject a gamma ray —a super x-ray. Or one of the protons may expel a positron and turn into a neutron. The escaping positive charge also carries away energy. In either case a stable nucleus results.

⊙

Modern Alchemy

The much-publicized "atom smashers" of the physicist are devices for shooting protons, deuterons, or other nuclear projectiles into various types of nuclei. The present-day physicist thus has achieved the ancient goal of the alchemist: transmutation of one element into another.

The individual steps of a transmutation process can release a certain amount of energy. I said above that the nucleus of a helium atom weighs four times as much as a proton. But this statement is only roughly true. If I could start with 1.0078 grams of hydrogen, and convert all the material into helium, I should have only 1.0000 gram of that product. The remainder, 0.0078 gram, or a trifle less than 1 percent, would have been converted into energy. By Einstein's equation, we get 0.0078 gram of energy, or 1.7×10^{11} calories. This heat is 5,000,000 times greater than that produced by ordinary combustion of 1 gram of hydrogen in the presence of 8 grams of oxygen.

There are several ways of building up helium from hydrogen. One of the simplest and most effective was suggested by H. Bethe. A normal carbon nucleus, with six protons and six neutrons, plays an important role in the process. This nucleus picks up protons, one at a time, and grows into different elements. At the moment the nucleus picks up its fourth proton, a helium nucleus splits off, leaving the carbon in its original form, ready to start the sequence over again. The action of carbon is similar to that of a catalyst in ordinary chemical reaction.

Table 7 details the various steps in this so-called carbon cycle. The first step is simple enough. The addition of a proton gives seven such particles in the nucleus of the product. Any atom with

Fig. 168.   General view of the electrostatic accelerator used in the Kellogg Radiation Laboratory at the California Institute of Technology for the study of nuclear reactions of astrophysical interest.

TABLE 7.   STEPS IN THE CARBON CYCLE BY WHICH HELIUM IS BUILT UP FROM HYDROGEN.

Step	Starting substance			Reaction	Product		
	Element	No. of protons	No. of neutrons		Element	No. of protons	No. of neutrons
1	C	6	6	Add proton	N	7	6
2	N	7	6	Emit positron	C	6	7
3	C	6	7	Add proton	N	7	7
4	N	7	7	Add proton	O	8	7
5	O	8	7	Emit positron	N	7	8
6	N	7	8	Add proton	C He	6 2	6 2

seven positive charges is nitrogen, irrespective of the number of neutrons it possesses. But the nitrogen so formed is unstable; the proton—or one of the protons—emits a positron and turns into a neutron. The cycle goes on until we have recovered our original carbon atom and a helium atom. The extra mass has been converted into energy, partly that of the emitted positrons and the rest in the form of gamma radiation.

The direct union of two protons, with emission of a positron to form a single deuteron, constitutes another, much simpler, process for the release of atomic energy. This reaction can go on also to build a helium nucleus, by the steps shown in Table 8. This thermonuclear process is related to those used in the H-bomb, which are currently being studied in the laboratory with the hope of harnessing them for peacetime use.

Which process, the carbon cycle or the proton-proton reaction, dominates in a given star depends on its internal temperature. The former demands the greater temperature and is much more sensitive to small differences in temperature. In the sun, we now believe, the proton-proton reaction probably produces most of the energy.

The great abundance of hydrogen in stars makes this element particularly important for stellar energy generation. If the sun consisted almost entirely of hydrogen, at its present rate of energy generation, its maximum life would be of the order of 10^{11} years. The time scale, for our own star, is thus comfortably long. For a giant star, like Capella, we are just about able to fit its observed brightness into the probable age of the universe. But a supergiant,

TABLE 8. DIRECT THERMONUCLEAR PROCESS
BY WHICH HELIUM CAN BE FORMED FROM HYDROGEN.

Step	Starting substance			Reaction	Product		
	Element	No. of protons	No. of neutrons		Element	No. of protons	No. of neutrons
1	H	1	0	⎰Add proton ⎱Emit positron	H	1	1
2	H	1	1	Add proton	He	2	1
3	He	2	1	⎰Add another ⎮ He³ ⎱Emit 2 protons	⎰He ⎱2 H	2 1	2 0

like Rigel or Deneb, would convert all of its hydrogen into helium in a few million years. We are thus forced to conclude that some stars may be of recent formation and even that the process of stellar birth is still going on.

Spitzer, Whipple, and Bok have theorized that certain small dark globules—now visible as minute black patches against the luminous background of bright nebulae—may be stars in the pre-natal state. The pressure of starlight gently squeezes the gaseous mass into a spherical form of dwindling size, until gravitation becomes strong enough to complete the job. This entire condensation process takes so long, however, that we can scarcely expect to witness it for even a single star in the course of human history.

The carbon cycle and the proton-proton reaction are atom-building processes. They suggest the possibility of a general evolution of the elements, and indeed atomic scientists have recently developed plausible theories for the formation of atoms heavier than helium. These reactions require regions of extremely high temperature and density. When Eddington originally suggested that suitable conditions might exist in stellar interiors, Jeans reasonably pointed out that the stars were not hot enough. Eddington gave a reply that is now classical. He told Jeans to "go and find a hotter place." As regards the average star, the evidence still seems to favor Jeans. But recent calculations indicate that once a star has burned all its hydrogen into helium, it must condense and grow hotter inside. Eventually it will become hot enough (about 100 million degrees) to fuse helium into the more complex nuclei of carbon, oxygen, and neon. When the helium is exhausted, the temperature again rises and a variety of other nuclear reactions result in the synthesis of the remaining elements.

$$\odot$$

Nuclear Fission

Radioactive atoms show two kinds of instability. There are the alpha, beta, and gamma emissions previously mentioned. Also there exists the possibility of atomic fission, produced by a neutron entering a nucleus that is on the verge of instability.

We find, in nature, two primary varieties of uranium atoms. Each has 92 protons. The more abundant isotope has 146 neutrons; thus there are 238 mass units in all. The isotope of lesser abundance has 143 neutrons, so that its total mass is 235 units. We call these isotopes U238 and U235, respectively.

When a neutron strikes a nucleus of U235, one of two things may happen. The neutron may stick, to form a new isotope; or the neutron may cause the nucleus to split into two unequal parts. This latter process releases an enormous amount of energy. Each fission also provides several new neutrons, which in turn can cause fission of other atoms. Thus, if fission were to start in a mass of pure U235 large enough that the ejected neutrons could not escape, these neutrons would in turn cause additional fissions, and a chain reaction would ensue. If the number of released neutrons is not controlled and multiplies indefinitely, the accumulated energy will cause the entire mass to explode. Calculations show that even if the reaction had an efficiency of only a few percent, each kilogram of U235 in a bomb would be equivalent to 300 tons of TNT.

The story of the development of the atomic bomb is too long for a detailed review here. There are, however, a few basic facts relevant to the character of atomic nuclei. Experiments showed that slow neutrons are more effective than fast ones in producing fission of U235. Also, as an alternative to splitting, U238 may absorb slow neutrons efficiently to make an atom of U239. This nucleus is unstable. Two of its neutrons eject electrons in rapid succession, to form neptunium (Np) and plutonium (Pu)—new chemical elements with numbers 93 and 94 in the periodic table. Further, plutonium behaves much like U235, fissioning after absorption of a slow neutron.

Of the two types of uranium, only U235 is readily fissionable. Separation of U235 from U238 is a difficult process. The two atoms are chemically similar, so that the methods available must depend upon the small difference in mass. Also U235 is much less abundant than the heavier isotope—natural uranium contains 140 atoms of U238 to one of U235. Plutonium, however, is a distinct element and chemical procedures make the separation of this substance from uranium a relatively easy matter.

Instead of working with U235 directly, the atomic scientists thus proceed as follows. First, they obtain pure uranium, a mixture of both isotopes. Lumps of this metal are embedded in a large "pile" of pure carbon, like raisins in a cake. The purpose of the carbon is to slow down the ejected neutrons, because the fission-produced neutrons are too fast for efficient working of the process.

The automatic splitting of U235 atoms produces neutrons. Some of these go into the U238 nuclei to form plutonium. The remainder strike other U235 nuclei and maintain the chain reaction. After a lapse of time, the metallic chunks are removed and the accumulated plutonium is extracted by chemical means.

In small amounts (exact size classified) the fissionable material, U235 or Pu, is not subject to spontaneous explosion. The dangerous particles, namely, the neutrons, leak completely out of the substance and are not captured. But assemble two or more of the small masses to form a large one that holds on to its atom-splitting neurons, and boom—Hiroshima!

The atomic fissions of plutonium and uranium 235 release vast amounts of energy. But processes of atom splitting seem much less likely to occur in normal stars than those of atom building. The former are of special importance because they first gave absolute confirmation of our basic hypotheses concerning the conversion of matter into energy.

$$\odot$$

Internal
Structure
of the Sun I have previously stated that the internal temperature of the sun is of the right order of magnitude for the proton-proton reaction. Just how do we infer the existence of such high temperatures in the solar interior? First of all, we know that the densities are higher inside than outside. In a pile of spring mattresses, the lower ones tend to be compressed more than the upper ones. In the same way, the outer layers of a star tend to compress the inner layers.

While pressure tends to compress, high temperature tends to expand the gas. Hence the resulting density at any point in the star depends on the magnitude of these two counteracting forces. We owe our knowledge of stellar interiors to a number of investigators: Lane, Emden, Eddington, Jeans, Milne, Russell, Chandrasekhar, and Schwarzschild, to mention only a few. Although the details of their calculations are too technical for presentation here, the reader can visualize the general procedure. We start at the solar surface, where we know the temperature and density. We choose a second point a little way below the surface and assume a temperature. The density follows immediately because we can compute the weight and pressure of the upper layer. We then take a third point, again assume a temperature, and calculate the density; and so on down

Fig. 169.   Sir Arthur Eddington, pioneer prospector of stellar interiors. (Clarke, Cambridge, England.)

to the center. At the end of such a calculation we have two checks. The total mass of all the layers must agree with the observed mass. And the calculated mean density must be 1.4, in accord with the measured value.

The actual procedure is not quite so much a matter of trial and error as the preceding paragraph suggests. Indeed, if we followed this course exactly as described, we should find not one but many temperature distributions that would fit. However, most of the energy generation probably occurs in the inner core. Thus the total flow of heat across any spherical surface whose center is the sun's center must be constant, and equal to the observed flow at the surface. This fact provides a further guide in the calculations.

The insulating power or opacity of the gaseous layers determines the rate of heat flow. Moreover, this insulating effect (or its inverse, conductivity) depends on both the density and the temperature of the gas. The necessity of accounting for this additional effect gives a unique determination. The latest calculations, made by M. Schwarzschild (1957), indicate that the central temperature is of the order of 15,000,000°C, with a central density about 130 times that of water.

The actual computations allow also for such refinements as change in average molecular weight along a solar radius. If the sun consisted of pure ionized hydrogen, each atom would contribute a mass of one unit and each electron effectively zero. Thus the average contribution would be one-half mass unit. Most of the uncertainties in the temperature calculations derive from our lack of precise knowledge of the actual chemical composition of the interior layers.

The composition usually assumed contains a somewhat higher proportion of the heavier elements than that given previously for the external layers. The relative abundances among the heavier elements are approximately those first determined by Russell for the solar atmosphere—the so-called *Russell mixture*. To these basic ingredients we add various fractions of hydrogen and helium until the computed opacity agrees with the observations.

Most calculations assume, in effect, a certain amount of gravitational separation. Some of the lighter hydrogen has risen to the top of the atmosphere. The calculations further imply that convection, that is, vertical mixing of the gases, tends to oppose further separation. If such stirring did not occur, the hydrogen layer would be less than 1000 kilometers thick. Heavy metals—even uranium —would compose most of the sun. Indeed Eddington, in his original calculations, assumed that the sun and stars consisted chiefly of iron. But he then found it impossible to match the observed and calculated opacities for the transfer of heat.

I do not seriously question the validity of the current assumptions about chemical composition and convection. The prominences are conclusive evidence of vertical mixing near the surface. And Unsöld has shown that the subsurface stratum, where the hydrogen atom undergoes ionization, must also be turbulent. I have referred to this phenomenon earlier (p. 217) in connection with granulation and spicules. When astronomers had to rely on the carbon cycle for generation of solar energy, their calculations indicated that the central core was highly convective. Now that they consider the proton-proton cycle responsible, theory shows that very little convection occurs in the core.

In general, there are three ways for transport of heat: conduction, convection, and radiation. Familiar examples of the three respective processes are: a hot spoon handle, a hot-air furnace, and an electric radiant heater. Conduction is essentially a phenomenon of solids and does not apply to the sun, which is gaseous throughout. Convection, wherein the hotter gases rise and the cooler fall, may carry some of the energy outward. But most astronomers now agree that radiation is the most effective method for transport of heat from level to level, except in the outer 10 percent or so of the solar radius, where convection does play a significant part.

One minor observation indicates the need for caution in regard to some of the foregoing assumptions. At the time of total eclipse, when the moon has obscured the photosphere, we have an opportunity to examine the structure of the outer layers. The flash spectra show very extended lines of hydrogen and helium, indicating the great height to which the H and He extend in the solar atmosphere. Lines of the more abundant metals are less extended; those of the so-called rare-earth elements are very compressed. One gets the impression that the heavy rare-earth atoms have partially settled to the lower depths of the atmosphere. However, the observed effect is more probably one of excitation or ionization than of gravitational separation.

The problem of rotation and general circulation of stellar interiors has not as yet been completely solved. The fact that surface layers rotate more rapidly at the equator than at higher latitudes suggests that the atmosphere is stirred from within. In other words, the rate of rotation increases with depth. Somewhere in the deep interior lies a key to the problem of sunspots and other variable atmospheric features.

12

The Sun and the Universe

The scientist must consider the sun also in its relation to other stars. The tiny points of light that adorn the heavens are suns: huge spheres of hot gas. In fact, the great majority of the stars visible to the naked eye are hotter and intrinsically brighter than our own sun. The faintness of these suns is due to their great distance. Light, traveling at the speed of 186,000 miles per second, requires only 8⅓ minutes to reach us from the sun. But it takes more than 4 years to reach the earth from the nearest star; in other words, we say that this star is more than 4 *light-years* away.

Astronomers regard any star lying nearer than 100 light years as a close neighbor. As justification for this view, I point out that the great Milky Way system, of which our sun is an insignificant member, is a flattened disk containing perhaps 100,000,000,000 stars within a diameter of 100,000 light years.

The structure of our stellar system scarcely comes within the subject matter of this book. [This subject is discussed by B. J. Bok and P. F. Bok, *The Milky Way* (Harvard University Press, Cambridge, 1957).] But let us examine some of the characteristics of the individual stars. For the nearest stars, at least, we can measure the distance directly. These objects show a minute displacement caused by the motion of the earth around the sun. If you close one eye and waggle your head, you will find the nearer objects appear to shift their positions with respect to the more distant. The magnitude of the shift is a measure of the distance. This is the method of parallaxes, more fully described in Chapter 1. In that description, the baseline was a distance measured on the earth. Here, however, because the stars are so far away, the baseline must be as long as possible, and we therefore use the diameter of the earth's orbit. Astronomers have also developed numerous indirect procedures for the determination of stellar distances, but these need not concern us here.

Any star's apparent brightness depends on its true luminosity, or candle power, and its distance. It is a simple matter to calculate luminosity when we know the distance and apparent brightness. The results show a wide variation in the true brightnesses of individual stars. The sun lies about midway between the two extremes. There are few known stars as much as 10,000 times brighter or 10,000 times fainter than the sun.

A star has one other easily measurable characteristic: its spectrum. The variety of spectral types is so great that early astronomers had difficulty in detecting any simple relation. Some stars, like our sun, exhibit predominately metallic spectra. Others, like Sirius, show the hydrogen lines as their main feature. Still others, like Antares or Betelgeuse, possess the banded spectra of molecules.

At first, scientists were inclined to attribute the varieties to actual difference in chemical composition, at least until they realized that certain types of spectra were always associated with stars of a given color. Then they noted the existence of stars with types intermediate between the main groups. Thus, Miss Cannon, at Harvard College Observatory, found that Secchi's original four types could be better fitted into a more detailed, regular sequence, as follows:

Type	Color	Chief spectral characteristic
O	Blue	Hydrogen, ionized helium
B	Blue-white	Hydrogen, helium
A	White	H, ionized metals
F	Straw	Ionized metals
G	Yellow	Ionized and neutral metals
K	Orange	Neutral metals
M	Red	Molecules

There are a few additional classes, R, N, S, and W, but so few stars fall into these groups that we need not consider them here.

The color sequence is one of temperature, ranging from the hot blue O's to the cool red M's. This interpretation is consistent with the spectra. For only in the hottest stellar atmospheres can the helium atoms be ionized; only in the coolest can molecules form.

A graph of the absolute magnitudes of the stars against their spectral characteristics is most illuminating. We term such a graph the Hertzsprung-Russell diagram, in recognition of the codiscoverers of this relation. On it, the majority of the stars fall into two principal groups: the so-called main and giant sequences (Fig. 170).

The former comprises the line stretching diagonally across the page. Our sun is a G star, with absolute magnitude 5. Five magnitudes correspond to a range of 100 times in brightness. The faintest star shown in the graph is of absolute magnitude 13, which signifies that the star is about 2000 times fainter than the sun.

The hotter the star, the brighter its surface per unit area. Thus, of two stars whose total brightness is equal, the hotter one will be the smaller. Among the stars of the giant sequence, as we proceed from B toward M in the direction of lower temperatures, we encounter increasing size. The giant M's are truly enormous. To emit as much light and heat as they do, some of these cool stars must have radii larger than that of the earth's orbit. If we were to replace our sun with one of these stars, we should find the earth and, in some instances, even Mars well below the "surface." Along the main sequence, which comprises the group running diagonally downward to the right, a reverse condition obtains. The red stars are both cool and small; the blue ones, hot and large.

The peculiar distribution of points on the Hertzsprung-Russell diagram is highly suggestive. The gaps, we must assume, indicate

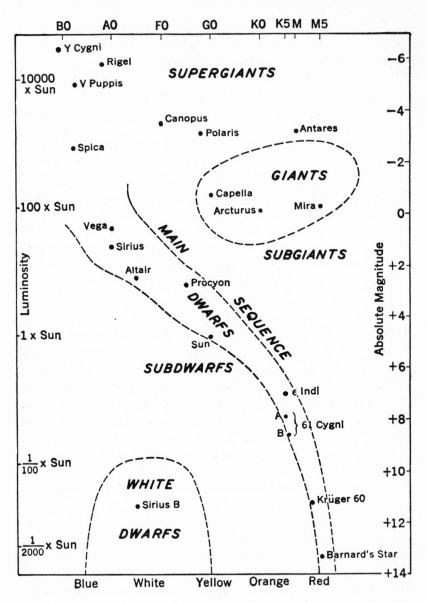

Fig. 170.   Schematic Hertzsprung-Russell diagram. Broken lines roughly define the regions within which stars occur in greatest abundance. The positions of a number of well-known stars are shown. (Adapted from Skilling and Richardson, *Astronomy,* Henry Holt & Co.)

forbidden zones. A star with properties that would put it in one of the gaps must be unstable, in the sense that its evolution would carry it quickly into one of the favored regions of the graph.

Double stars—whirling pairs bound together by gravitation—give us some clue to the nature of these forces. Hence we can determine the masses of such stars by an extension of the reasoning proposed in Chapter 2 for measuring the mass of the sun. We find that the most luminous stars are, in general, the most massive.

The horizontal coordinate in Fig. 170, spectral type, is primarily a measure of surface temperature. For a given star, the total energy output, indicated by the vertical coordinate, is equal to the surface brightness multiplied by the area of the star. Since the temperature gives us a measure of the surface brightness, we readily calculate the size of the star. The Hertzsprung-Russell diagram, therefore, means that main-sequence stars of a given mass must tend to possess about the same radius. Among the giants we find no such simple relation.

Astronomers have attempted to interpret the diagram in terms of stellar evolution. As Herschel so picturesquely indicated, our position with respect to stellar evolution is like that of a man who walks for a day in the forest. He may not see a single leaf unfold. But on all sides he notes sprouting seeds, saplings, sturdy oaks, decaying stumps. And from such evidence he might deduce a fair picture of the life history of a tree. Similarly astronomers have attempted to infer the life history of a star.

According to our first guess, the bloated red giants were young stars. Their average densities are low—less than that of the earth's atmosphere. They possess large masses. It was natural to conclude that compression caused these giants to contract, the while growing hotter, and moving left on the diagram, until further compression at the center was impossible because of the high densities. Thereafter, these stars were supposed to slide down the main sequence, step by step, toward ultimate extinction.

This picture for years seemed attractive, but quantitative studies of conditions in stellar interiors, and increasing knowledge of nuclear reactions, have revealed several flaws. For one thing, we could find no compression limit. The high densities of stellar interiors would crush the atoms like eggshells. The discovery of a third group of stars—the white dwarfs—confirmed the idea that compression had by no means reached a limit in stars of the main sequence. The companion of Sirius, the first white dwarf discovered, is so dense that a teaspoonful of its material would weigh approximately 1 ton. The familiar atoms of our experience are, in that star, squashed beyond all recognition.

A second difficulty lay in the relatively long time scale required for stars to pass through their life cycles, especially in the latter stages of the main sequence. I have said that studies of radioactivity and general chemical composition of meteorites and the earth have indicated an age of 4500 million years, nothing like enough time to evolve the red dwarfs by the foregoing picture. Of course, our earth may conceivably be a mere infant in this universe of stars. But another, somewhat elusive bit of evidence points toward the conclusion that the stars cannot be much older than the earth.

Fig. 171.   Spiral galaxy in Ursa Major. (Lick Observatory.)

In the sky, we find numerous Milky Way systems very much like our own. Each of these so-called *external galaxies* consists of thousands of millions of stars, usually arranged in the form of a flattened, rotating disk. Many exhibit a spiral structure. The nearest of the groups is about 1,000,000 light years distant; the farthest yet surveyed are several hundred times more remote. The amazing fact is that these galaxies appear to be receding from us. Their spectral lines are shifted toward the red. The fainter and more distant the object, the greater this shift, and hence the greater its speed of recession. [See Harlow Shapley, *Galaxies* (Harvard University Press, Cambridge, 1943)]. Red shifts have recently been measured corresponding to velocities as great as 38,000 miles per second, or about one-fifth the speed of light.

Although such velocities may seem fantastic and incredible,

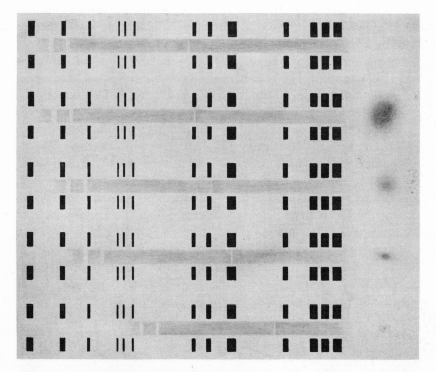

Fig. 172.   Red shift of spectra of galaxies, schematic. The top row shows an unshifted spectrum, relative to laboratory standards along each side of the spectrum of the object. The remaining spectra of progressively more distant galaxies show greater and greater shift toward the red (right).

scientists have found no other explanation for the observed red shifts that is consistent with modern physics. We are forced to conclude that the universe is expanding—as if from an explosion that occurred several thousand million years ago.

According to Einstein's general theory of relativity, however, this is precisely what one would expect. The theory indicates that, if we could travel backward in time, we would find the matter of the universe in a more and more compressed state. And, calculating backward from the red shift, scientists find that 7 or 8 thousand million years ago the universe began to expand from a state of maximum possible compression. The density then was probably greater than a million million (10^{12}) times that of lead; a teaspoonful would have weighed over a million tons.

Einstein's theory thus predicts that none of the objects we see in the sky can be older than this 7-8 billion years. In such a lifetime, stars less massive than the sun, in particular the red dwarfs, cannot have changed much. We must therefore abandon the early picture —red dwarfs are not old and dying stars.

On the other hand, stars a little more massive than the sun must have changed appreciably. And the very massive stars, the supergiants like Rigel and Deneb, will change drastically, beyond all recognition, in a much shorter time.

The arguments outlined above, and others outside the scope of this book [see Bok and Bok, *The Milky Way;* also Cecilia Payne-Gaposchkin, *Stars in the Making* (Harvard University Press, Cambridge, 1952)], have recently led A. Sandage, M. Schwarzschild, and others to propose an entirely different evolutionary interpretation of the Hertzsprung-Russell diagram. As a star condenses from a primeval cloud, it moves quickly to the left on the H-R diagram, radiating away its energy of gravitational contraction, to a position on the main sequence. Its particular position will be determined primarily by its mass. According to the modern picture, a star spends the major part of its life on the main sequence. For stars of solar mass or less, this major part will be many thousand million years. But the vast majority of red giants, instead of being young stars, are now believed to be old stars, running short of hydrogen fuel.

The red giants have long puzzled astronomers. For main-sequence stars, the proton-proton reaction and the carbon cycle seem entirely adequate as energy sources, the former predominating in the sun and less massive stars, the latter in the larger and bluer stars. But when we studied the red giants, we found that, if they were built on the same model as main sequence stars, their centers would be too cool for either of the hydrogen-burning nuclear reactions to occur. Then where could they get their enormous energies? We have only recently begun to feel sure of the answer: red giants have an internal structure entirely different from that of main-sequence stars.

The main-sequence star is fusing hydrogen into helium, at a rate per unit mass that is faster the brighter the star. Gradually the hydrogen is consumed until the core is composed entirely of helium and the original sprinkling of heavier elements. But outside the core, the star has still a substantial amount of hydrogen. Energy will be generated in a hydrogen-burning shell that surrounds the helium core. During the growth of such a shell, the star increases its brightness; the outer layers are distended into the tenuous atmosphere characteristic of a giant star. The radius and the luminosity both increase, but in such a proportion that the brightness per unit area decreases and the color grows redder. By this process the ordinary main-sequence star evolves into a red giant, moving up and to the right in the Hertzsprung-Russell diagram.

As the hydrogen in the shell is consumed, the helium core grows larger and hotter. When its temperature rises above a hundred million degrees, the helium begins to fuse into heavier elements, such as carbon, oxygen, and neon. The star brightens further, and moves into the region of red supergiants. What happens next is a problem that astronomers have not yet solved. However, the end seems fairly clear, even if the particular evolutionary steps are obscure. When a star has consumed all its sources of nuclear energy, it will end its life as a superdense white dwarf, in the lower left-hand corner of the H-R diagram.

Our brief study of the sun has led us far afield, from the deep interior into the most distant reaches of the universe. Our journey has not been without profit, for, as we return to the sun, we have learned that it is not unique among the stars. There are many stars that have similar characteristics.

Each star is a great nuclear furnace—an engine for converting matter into energy. Each second our sun changes about 4,000,000 tons into radiation and scatters it in space. By terrestrial standards, 4,000,000 tons is a lot of matter. In mass, it is equivalent to about

60 ships the size of the *Queen Mary*. In energy, it could supply the entire present level of power needs of the earth for 2,000,000 years.

Many stars show marked variability in both the intensity and the color of their light. The variables of greatest range occur in the giant sequence. Our sun, as we have seen, appears to be unusually constant. Despite the fact of the sunspot cycle, its radiation in the visible range seems to have varied by less than 2 percent in the 30 years we have measured it. There is every reason to believe, however, that changes in the far ultraviolet may amount to several hundred percent. But I shall leave this aspect of solar variability to Chapter 14.

⊙

The Ice Ages

The question has often been raised whether solar variability may not have been greater at certain periods in the past. Geological records show unmistakable evidence of several extended glacial epochs. At first sight these might be thought to represent times when the solar atomic fires were banked.

As we survey the stars for evidence of such variation, we do find a few, such as R Corona Borealis, that turn off the heat at irregular intervals. But these seem to be giant stars, totally unlike our sun.

There is, in any case, serious doubt whether a cooler sun could cause ice ages. Remember the name, "ice age," not just "cold age." These were times when great sheets of ice, perhaps thousands of feet in depth, covered the continent of North America as far south as New Jersey, Ohio, Indiana, Illinois, and the Missouri River Valley. The slow advance of the enormous ice masses carved out the beds of the Great Lakes, scraped the gashes now known as the Finger Lakes. When a period of warm dryness sent the ice into recession, it left behind huge masses of debris.

G. C. Simpson was the first to argue that if the sun were to turn down its heat, an ice age would not result. Evaporation would be diminished. The winters would be colder, but there would also be less snow. The heat of the summer would also be less, but we should lose less of it in the melting of the snow. In fact, we are faced with the paradox that a lessening of solar heat might even cause our present polar snows to retreat.

Simpson further suggested that the converse of this paradox may also be true. A small increase in the sun's heat would enhance evaporation from the oceans. The summer would not suffice to melt all the additional accumulated snow. The ice would slowly advance southward. With a further increase in solar heat we should expect most of the precipitation as rain and no difficulty in melting the winter snows.

While these arguments have appeared to me more convincing than their converse, many scientists have felt an understandable difficulty in accepting any such "hot-sun, cold-earth" theory of the ice ages. B. Bell has recently proposed certain modifications in the simple theory which may lessen this difficulty. She suggests that an ice age must be preceded by a long period of diminished solar radiation. And indeed, geologists have found evidence of an increasingly cold and dry climate through many millions of years preceding the latest or Pleistocene ice ages. In such a period the oceans would grow cooler, reducing the heating effect in high latitudes of warm currents like the Gulf Stream, and thus permit the formation of permanent ice in the polar seas—a condition which exists at present, but which does not seem to be typical of the earth's past. For example, incredible as it may now seem, through the major part of geological history the climate of Greenland has been more suited to the growth of magnolias than of ice sheets.

The earth, especially the oceans in high latitudes, must therefore be precooled in preparation for the growth of an ice sheet. Then as the solar radiation rises toward normal, increasing the evaporation and precipitation, ice sheets grow and flourish as described by Simpson. As the solar heat continues to rise, the summer melting eventually predominates and the ice sheets recede and in time disappear. If the solar radiation declined again to the level where it had been when the ice sheets formed, no ice sheets would appear because the earth had not been precooled. Thus Bell suggests that ice ages may result from a rise in the solar radiation, but only if this rise occurs within a period of prolonged depression of the solar radiation.

The theories I have so far described assume a change in the total solar radiation, that is, in the solar constant. You may well ask at this point: do scientists know whether or not such changes can take place, or actually have taken place, in the solar radiation? And I must admit that we cannot yet answer this question with any degree of certainty.

On the other hand, as I have previously mentioned, we are certain that large variations do occur in the far ultraviolet and in the corpuscular radiation from the sun. H. C. Willett, a meteorologist at Massachusetts Institute of Technology, has suggested that variations in these radiations may be what determines whether we shall have glacial or nonglacial climate—by influencing the patterns of atmospheric circulation rather than by an over-all warming or cooling of the earth. While the why and how of it are not yet understood, a growing body of evidence suggests that variations in the solar ultraviolet and corpuscular radiations do indeed influence our weather. Willett argues, moreover, that all the various cycles and subcycles, from ice ages to the recent warming of the winters in high latitudes and the world-wide recession of glaciers, differ only in duration and amplitude; and his theory proposes for all of them a common explanation.

The ice ages are extremely complicated phenomena, however, and may well have several contributing causes. The two postulated types of solar cause that I have discussed may in fact well supplement each other. The foregoing theories need not be mutually exclusive and should not be so regarded.

As an added complication, geological evidence indicates that not all glaciation has been polar. Around 250 million years ago, vast ice sheets occurred in regions which form modern India. To account for such vagaries of ice distribution, some geologists have suggested that these now equatorial regions must have been in the vicinity of a pole when they were glaciated. This explanation states either that the continents have drifted—like slices of bread floating on a sea of molasses—or that the earth has shifted bodily relative to its axis of rotation. Such liberties with the earth have long excited lively controversy. However, recent arguments, by T. Gold, seem to indicate convincingly that the earth not only can shift relative to its axis of rotation, but must do so in such a way that its moment of inertia will remain a maximum. In other words, as mountain ranges are slowly uplifted, and others are eroded, the whole earth, also slowly, will tend to shift so as to keep the largest masses of high mountains near the equator.

But the most recent glacial period occurred only 20,000 years ago, when the axis and continents must have been in substantially their present positions. The existence of such motions, therefore, can only explain how some unlikely parts of the earth were glaciated instead of other more apparently likely regions. It does not explain why glaciation occurs at a given time in geological history, or indeed why it occurs at all.

I have said that there may be many contributing causes for the ice ages. One should consider possible effects of changes in the composition of the earth's atmosphere. Certain gases, such as water vapor and carbon dioxide, have a pronounced "greenhouse effect" —they trap infrared radiation and reduce the escape of heat from the earth into outer space. Also one must consider the possible consequences of an encounter between the solar system and a gaseous nebula, or cloud of cosmic dust. These objects are fairly common in space but their density is far too low to have any significant screening effect on direct sunlight. There is another possibility, however. Experimental work by V. J. Shaefer, of General Electric Company, has demonstrated the importance of minute particles in the air, to act as *condensation nuclei* or centers around which water vapor will readily form into droplets. Artificial rainmaking consists essentially in providing extra condensation nuclei for the formation of raindrops. It is similarly possible that extra condensation nuclei from natural sources—such as cosmic dust clouds or unusually great volcanic activity—may play a role in the ice ages by greatly increasing the precipitation. Of the two, volcanos look more promising. The density of cosmic dust seems too low to be effective.

It would be interesting to dwell further upon this problem, but space forbids. [For more detailed discussion, see Harlow Shapley, ed., *Climatic Change* (Harvard University Press, Cambridge, 1954).]

⊙

Origin of the Solar System

One final question I should like to raise before closing this chapter: is it possible to consider the sun as a parent of the earth? The origin of the solar system is, like all events so long ago, shrouded in mystery. Many hypotheses have been proposed. In most of them critics have found serious flaws.

On the so-called "nebular hypothesis," first put forward by the

French mathematician Laplace in the early 19th century and later elaborated and modified by others, the sun is supposed to have shed successive rings of matter during its own formation by contraction from a primordial nebula. As a rotating body contracts, it spins faster and faster, until it becomes unstable and leaves behind it equatorial rings of matter. These rings fragment and the pieces coalesce into planets. Laplace's theory thus makes the birth of planets a natural consequence of stellar evolution, and this is one of its most attractive features, since it would imply that many stars may have planets, and in no way suggests that the earth is unique.

The contemporary German philosopher and physicist, F. von Weizsäcker, has also elaborated a theory that has the sun and planets forming out of a rotating primeval nebula. While Laplace's nebula rotates smoothly on its axis, von Weizsäcker's is seething and churning. The violently turbulent motions create regions of high density, some of which manage to condense into planets, while the remainder of the gas falls toward the center and forms the sun.

However, critics have found serious difficulties with both of these theories. If the sun had been formed in either of these ways, it should now be spinning much more rapidly than it does. Or, to put it differently, working backward from the present observed rate of rotation, we find that the sun would never have rotated rapidly enough to give off rings or to sustain violent motions. Scientists also question whether either the rings or lumps of gas would actually condense into planets.

An entirely different variety of theory, advocated by Jeans and others, would make the birth of planets a very rare occurrence, since it requires a collision or near collision between our sun and another star or stars. The encounter would fill space with debris, pulled from the sun or ejected from it by eruptive processes. But there are two apparently fatal objections to all theories of this type. The expelled gases would be far too hot to coalesce into planets. And even if planets could somehow form, their orbits would be much smaller and more elliptical than those we observe.

A third type of hypothesis relates the origin of the solar system to the formation of other systems such as stars and galaxies, and to the expansion of the universe. According to the theory of relativity, at the moment when the universe began to expand, its density was so high that the galaxies, stars, planets, even atoms, could not have existed as individuals. Instead the very dense matter filled space more or less evenly. According to a theory recently put forward by D. Layzer, the subsequent evolution proceeds as follows. Minor irregularities of various sizes develop in the primordial matter. The expansion facilitates condensation into separate cloudlike masses. Each cosmic cloud has a complex hierarchic structure, each consisting of smaller clouds which in turn consist of still smaller cloudlets. The largest cosmic clouds become galaxies and clusters of galaxies; smaller ones become clusters of stars; and still smaller ones develop into objects like our solar system. The primeval solar system, at the moment when it first separates out, contains the condensations (yet smaller cloudlets) that later evolve into the planets and their satellites. As the condensations solidify into planets and satellites, their orbits become more nearly circular as a result of frictional forces, which also tend to pull the planetary orbits into a common plane.

Layzer's theory appears promising and, at least as I write, has received no serious criticisms. I like it because it makes planets as normal as stars and galaxies.

I must caution the reader, however, against accepting without reservation any theories about the state of the universe in the very distant past. There are too many unknown factors. Perhaps, in the early evolutionary stages of the universe, the forces of electromagnetism may have been comparable to those of gravitation. Neglect of these forces may be one reason why we find it so difficult to reconstruct the details of the evolutionary process. In any case, the further we extrapolate, either backward or forward, from present conditions that we can actually observe, the more cautious we should be.

Keeping this reservation in mind, let us see what we can predict for the future of our sun. Our sun is a typical medium-sized main-sequence star. It should continue to shine steadily at something like its present brightness for tens or hundreds of millions of years, converting hydrogen to helium in its core. But eventually the hydrogen will approach exhaustion in the core, and begin to burn in a shell around the core. As this happens the sun will grow brighter and larger and redder. (Indeed, Schwarzschild has estimated that the gradually changing composition of the core has already caused a brightening of about half a magnitude over the past two billion years.) Slowly over the coming millions of years, the sun will swell to many times its present diameter and will shine with many times its present brightness. On earth, life will perish and the oceans boil dry in the greatly increased heat. Nor can we exclude the possibility that the earth itself may perish, swallowed up in the tremendously distended and perhaps prominencelike atmosphere of the sun.

Still later, when the hydrogen and helium both approach a state of exhaustion, the sun will cool and collapse to form a white dwarf. And finally it will fade to invisibility—a chilled clinker, still attended by such planets as survived its giant stage, planets to which it once gave light and life.

We cannot completely exclude the possibility of accident. Stars sometimes explode—and they may experience a collision. But even if such a catastrophe should occur, it would merely interrupt momentarily the path of evolution to ultimate and inevitable extinction. We are living in the age of cosmic fireworks. Indeed, this is the only age in which life is possible. For there is no evidence of any process that can come along and wind up the dead universe, to start the display over again.

13

Solar Power and Human Needs

Many of the topics detailed in prior chapters—the distance, temperature, chemical composition of the sun--may seem utterly remote from the world of human life. And I do not wish to imply here that the study of astronomy should require the justification of practicality. The primary value of all science is advancement of knowledge. Whether the results seem to have immediate or long-range applicability to human problems should be immaterial. However, history witnesses the fact that pure science often does bear practical fruit on the most unlikely branches. So it has been with solar research.

I have referred earlier to the slow but spectacular demise of the sun, as the core exhausts its source of atomic energy. But we are fairly sure that the sun will not brighten enough to destroy life on earth for some millions of years to come. The astronomer is accustomed to deal with long-range problems, but the questions I wish to raise here are not *that* long in range. We are more concerned with the immediate relations between the sun and the earth.

⊙

Our present-day civilization is using up the great natural sources of energy—coal, oil, and natural gas—far more rapidly than they are being replenished. The coal and oil deposits required long eras of geologic time to build. When they are gone we shall have to wait millions of years for natural processes to restore them. Will our motor cars and the great machines of industry have to stand idle in the meanwhile?

I have already pointed out that the energy in coal and petroleum came originally from the sun in the form of light. The great forests that existed in the carboniferous period, two or three hundred million years ago, collected sunlight and stored the energy in the wood. Animal and vegetable life flourished in great swamps that covered far more of the earth's surface than they do today. As ages passed, the trees decayed into peat bogs and were buried under slowly accumulating dust, sand, and mud. The weight of the overlying formations pressed the decayed vegetable matter into solid layers. Heat and pressure caused chemical changes, and gradually the great coal deposits of today came into existence. The accumulation of microscopic organic matter in the bottoms of other prehistoric bogs suffered a similar fate, to form petroleum and natural gas.

We are using up this fossil sunlight at an enormous rate. The present annual rate of coal mining in the United States amounts to about 550,000,000 tons. Oil production reached 2,700,000,000 barrels per year in 1957, while that of natural gas was a record 11 trillion (11×10^{12}) cubic feet. And consumption is rising year by year!

A survey made early in this century suggested that our major coal veins at the present rate of mining would last about 3000 years. The most recent survey, however, is less optimistic. We now find the reserves are much less than the early survey estimated. Also the percentage of coal that can be economically mined is much less than expected. Revised estimates indicate that our coal reserves will approach exhaustion within a hundred years. The discovery of new deposits or the mining of narrow veins of inferior quality may extend the period for a few more decades at most.

During the past quarter century, the oil reserves known at any moment could supply our country's needs for only 15 or 20 years. Fortunately, we are still finding new oil fields at a rate greater than we are using the product—indeed, as I write the oil industry is plagued with a surplus—and this implies that the number of undiscovered fields must still be very large. But we cannot expect to continue this process indefinitely. Current forecasts, taking account of the rising known reserves as well as rising demand, indicate that our oil and gas, like coal, will be exhausted in less than a century.

The distillation of oil from shale, as the Scots and Swedes now do commercially, may enable us to eke out our petroleum supply for additional decades. Also one can manufacture the equivalent of petroleum products from bituminous coal and lignite. But the cost of production will be considerably higher than at present.

For years the trend in the automotive industry of the United States has been toward larger cars and higher-powered motors. Such automobiles, in the interest of speed, comfort, and prestige, consume far more gasoline per mile than is necessary for adequate transportation. Sometime, we are likely to witness a reversal of this trend. Cars will become smaller and engines lighter, like today's popular European models. Large cars will be ever more a luxury, as our petroleum resources dwindle. The earlier we turn to economy of this sort, and avoid present waste, the longer will our resources last. Eventually we may expect government rationing of petroleum products, if the rising cost alone does not force restrictions on the consumer.

The foregoing figures on the duration of our reserves may well prove underestimates. Moreover, the picture of reserves outside the United States is somewhat brighter, promising some 300 years of production at present rates, so that we may expect to import far more oil in coming decades than we do now. But competition for this oil will intensify, with the increasing industrialization of the world as a whole, and will reduce the percentage available to us. Our exports, particularly of coal, may well increase. For example, we are already shipping coal to Newcastle! And some authorities claim that future coal production in Britain will be small indeed.

But, whether the exhaustion of supply will occur in a hundred or even a thousand years, have we not some responsibility to our descendants living on the earth at that time? The farmer plants trees that will bear fruit only for his children and grandchildren. National planning, likewise, often covers long ranges. We have large programs of reforestation, of dust control, of water-power development. The exhaustion of fossil fuels—if not prepared for— would be a world catastrophe. The problem is great. Perhaps we shall require hundreds of years for its complete solution. It is not too soon to make a beginning.

You may not realize to what an extent fossil fuels enter into your life. But look at the objects around you. Many of them are metallic. Others are plastic. Even your rugs, your furniture, your electric lights, your fabrics—in all probability coal, oil or gas entered somewhere into their production, either as a raw material or as a source of power in their manufacture. Electricity itself, to run your appliances, is generated partly by water power but mostly by fossil fuels.

The apparent abundance and wide distribution of these fuels are deceptive. Remember that we are squandering energy accumulated over millions of years in a few hundred years. No man who spends his inheritance a thousand times more rapidly than his father accumulated it can long remain solvent.

Perhaps I am unduly pessimistic about the extent of our reserves. New oil and coal deposits may well be far more extensive than geologists currently estimate. But it is not too soon to start a research program for the development of new liquid fuels based on sources other than coal or petroleum.

⊙

Other Sources of Energy

Mankind is faced with bankruptcy in its familiar energy sources. Where shall we look for energy when the coal, oil, and gas are gone? To wind and water? The energy in both of these comes to us secondhand from the sun. Unequal solar heating is responsible for the atmospheric motions. And greater use of wind power many not be as far off as supposed. Windmills have long been used to pump water on isolated farms throughout the world. Some experimental projects, involving studies of more efficient wind turbines, are already under way. A wind of 25 miles per hour, which is a fairly stiff breeze, striking against a turbine 100 feet in diameter, is capable of generating about 100 kilowatts of power. However, the cost of such an installation, compared with that of a hydroelectric plant of equal power, would be high.

With other sources of energy dwindling, we may look for further developments of water power. But in the United States the best sources have already been tapped. At that, water power furnishes only a scant 4 percent of our energy requirements. The problem of transmitting electricity great distances without severe losses is still unsolved. There are many small streams, each capable of delivering a few hundreds of kilowatts, but the distribution problems would make general development costly.

Both wind and water are relatively inefficient converters of solar radiation—and in different ways. More solar energy goes into heating the atmosphere than into the evaporation of moisture. But the natural tendency of watersheds to collect the rain and melting snows into a single usable stream more than compensates for the smaller fraction of sunlight used in the evaporation process. In most localities winds are fitful. High, steady winds usually occur only in mountainous regions, where the slopes and ranges tend to deflect air currents into a regular streamline path. With the disappearance of cheap fossil fuels, we may see a migration of some industries to the heights where the wind and water power are most plentiful.

To some extent, the development of atomic power from nuclear fission will postpone our fuel shortages. As I described in Chapter 11, two basic varieties of atomic reaction exist: fission and fusion. Uranium is the best-known example of the former. When a neutron happens to enter the nucleus of a uranium 235 atom, it causes that nucleus to split into two unequal fragments, with the emission of energy and also the release of more neutrons which can cause other U235 atoms to fission. If this process is carried out inside a container with special metallic rods to absorb the excess neutrons, the rate of fission can be accurately controlled. Remove those rods and the uranium explodes—as in the original atomic bomb.

From the very first, scientists have found it simple to control the release of radiation from uranium. The container remains relatively cool and shows no tendency to melt. Thus nuclear reactors for development of electric power and for operation of ships and submarines are relatively simple and a number are in operation today.

Atomic power is a rapidly expanding field. The British, who have assumed the lead in industrial applications of nuclear energy, have set a goal of 6,000,000 kilowatts of electric power, produced from uranium, by 1965. The French have set an even higher goal of 8,000,000 kilowatts by 1975. Japan aims at producing 3,000,000 kilowatts by 1965. The Russians have stated that they will be producing 2,000,000 kilowatts by 1960.

But, efficient as nuclear reactions are, we should have to consume annually 200 tons of uranium 235 to equal the coal consumption in the United States alone. This quantity corresponds to 30,000 tons of pure uranium, since the 235 isotope is present to only one part in 140 of the whole. Large amounts of the mineral are thus involved. Since available ores do not average better than 0.5 percent, we are talking about mining and processing at least 6,000,000 tons of ore per year.

In 1958, Jesse C. Johnson, Director of the Division of Raw Materials of the Atomic Energy Commission, said: "This country's present ore reserves of 70,000,000 tons represent only a ten-year supply at the (estimated) 1959 production rate." Thus uranium, like coal and oil, will be unable to supply for long mankind's ever more insatiable demand for power.

Of course we may, in the remaining years of grace, learn how to control the proton-proton reaction. Hydrogen—the fuel—is cheap and plentiful. And other light atoms, such as lithium, beryllium and boron, are also potential sources of fusion energy.

However, the fusion process, unlike the fission, is inherently difficult to control. In the proton-proton reaction, for example, two hydrogen nuclei must hit each other with sufficient force to coalesce and form a deuteron, with a release of energy. But the atoms will fuse only if they collide with speeds in excess of 100 miles per second. For the atoms to move this fast the gas must be heated to a temperature of about 10,000,000°F! An ordinary uranium bomb, fired in an atmosphere of hydrogen, will cause an explosion much as a dynamite cap will detonate a stick of dynamite.

To get a controlled hydrogen reaction, however, we must find a container capable of withstanding 10,000,000°F. No material substance will even come near to doing this. But a very strong magnetic field may possibly do it—for at least a minute fraction of a

Fig. 173.   Dr. Marvin Fox, of the Brookhaven National Laboratory, examines the wooden model of the research reactor, or atomic pile. The left-hand face of the cube has large round ports from which neutrons can be let out for scientific study. The right-hand face has numerous rows of holes, which represent the positions of uranium fuel cartridges in the graphite moderator. A sample of material in a container can be inserted into the reactor for bombardment by neutrons. It is quickly returned, now radioactive, for use in some experiment. Slow neutrons can be permitted to "leak" out of a graphite tube at the top. A shield of heavy concrete, 5 feet thick, encloses the entire reactor.

second. Newspaper releases as I write show that significant progress has been made, that temperatures of a few million degrees have been attained for a millionth or so of a second. At best, however, it will be many years before the process is made self-sustaining and practical, and even then it may require a machine that will be too costly in construction and too dangerous for any widespread use as a practical energy source.

Wide-scale research in the field of nuclear reactions—both fission and fusion—must be continued with government support. And the aim of the research should be cheap, abundant, and manageable atomic power—not bigger and deadlier bombs. We have a race

against time, with only 100 years or so to go, and with large benefits especially to fuel-poor lands from an early solution to the problem, since uranium can be transported more cheaply than the
bulkier conventional fuels.

If the controlled fusion of light elements proves impractical, as
well it may, we shall have to fall back eventually on direct solar
energy. Why use solar energy secondhand? Why wait for water of
the oceans to evaporate? Why wait for the wind to blow? Why not
tap sunlight directly?

We have already seen that the sun does send an enormous
amount of heat earthward. At noon, in the latitudes of the United
States, under ideal conditions, solar radiation falling upon 1 square
yard of surface is capable of yielding about 1 horsepower. If you
are thinking about some gadget fixed, let us say, to a 60-horse-
power automobile, to run it from the sun, visualize first of all a
great rectangular sail, some 20 feet by 30 feet, mounted flat on the
top of your car. This is the minimum size for a fully efficient engine.
When we multiply the area of the canopy by a factor of ten, to
allow for the general inefficiency of solar engines, we note how impractical the idea is in this form.

But 1 horsepower per square yard is about 3,000,000 horsepower
per square mile. At the rate of only 1 cent per kilowatt hour, one
day's accumulation would be worth about $200,000. An area
some 200 miles square receives enough solar energy to supply the
entire world with the equivalent of fuel at the present rate of consumption. Sunlight is clearly our greatest natural resource.

Let me make the picture still more graphic. The earth receives
enough heat from the sun in the course of a year to melt a layer of
ice 114 feet thick over the entire surface. Of course the melting at
the equator will be far in excess of this amount, while that at the
poles will fall to a relatively low figure. At that, only about one
part in two billions ($\frac{1}{2} \times 10^{-9}$) of the total solar energy reaches us.
The earth, from the standpoint of sunlight reception in interplan-
etary space, is like a 25-cent piece lying somewhere in a circular
field 1 mile in diameter.

But the energy that never reaches the earth is the least of our
worries. Our problem is to convert even a tiny fraction of the available radiation into useful form. There are numerous methods that
actually do work; unfortunately, they are either terribly inefficient
or extremely costly.

All heat engines—steam, gasoline, diesel, or what not—depend for their action on a simple principle of physics. The expanding gas that makes the engine run must start its cycle with a high temperature and end with a low temperature. The percentage efficiency of the engine is proportional to the difference in temperature of the initial and final products. The greater the difference, the more work the engine will perform and the greater will be its efficiency. To increase the performance of a steam engine, we superheat the steam under high pressure and discharge it into a condenser cooled with water.

The trouble is that solar radiation does not tend to produce extremely high temperatures. A piece of metal exposed to sunlight may get too hot to touch, but still not hot enough for operation of an efficient engine.

Many houses in proverbially sunny areas, like Florida, California, and Texas, use a bank of pipes exposed to sunlight as a source of hot water. However, unless great storage volume is provided a succession of cloudy days and nights will cause the owner to wish he were under the influence of a more beneficent sky!

There are many compounds that vaporize more readily than water, which we could utilize as the working substance for solar engines. Nevertheless, until someone invents a cheap and simple method for circulating the fluid and exposing it to solar radiation, the method will not be practical.

One can obtain a higher temperature by using mirrors or lenses to concentrate the radiation. The cost of such an installation, which must have motors to move the mirrors, makes it impractical today. Yet the plan does work. Years ago, at Mount Wilson Observatory, Abbot built a solar cooker, wherein a cylindrical reflector heated an oil to a high temperature; and other scientists produced a solar furnace whose temperature over a limited area was great enough to melt iron. More recently several successful solar furnaces have been built. The French operate a very large one in the Pyrenees. Another in the United States has been used for smelting refractory materials to produce a very high-quality ceramic. The sun is a very clean source of heat. Materials can be melted in the solar furnace without contamination by coal, oil, or any products of combustion.

Perhaps the greatest difficulty with direct solar energy is the problem of storage. The sun shines intermittently, whereas the

Fig. 174.   C. G. Abbot with toy solar cooker. The oven heats in 1 hour and bakes gingerbread in 35 minutes.

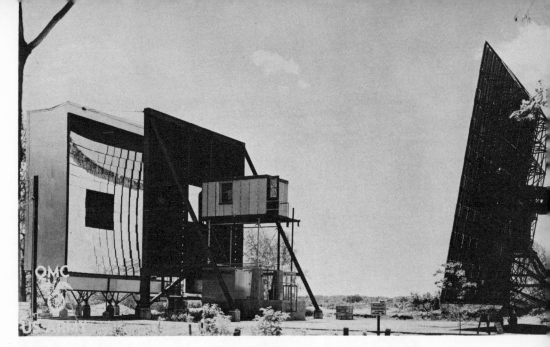

Fig. 175. Solar furnace operated by the U. S. Army Quartermaster Research and Engineering Center. The heliostat (*right*) measures 40 by 36 feet, bears 356 mirrors, and tracks the sun to catch light and pass it to the concentrator (*left*) through the lateral vanes of the attenuator. The 180 mirrors of the concentrator converge the sunlight into a 4-inch circle within the test chamber, where the temperature reaches 5000° F. (U. S. Army photograph.)

users' need is continuous and is often greater when the sun is not shining. We must save the energy collected in the day for use at night. Or, what is even more difficult, we must conserve the summer's excess radiation for winter operation.

Thus despite the fact that the sunlight falling on a 1-acre farm is worth at least $300 per day, the practical aspects of cashing in on the value appear at present insurmountable. I should not be so rash as to declare that a practical solar heat engine will not be developed. We may have to invent them and use them regularly. To be successful, such an engine must have negligible operating costs, however expensive the initial investment might be.

Another possibility is based on the fact that certain chemicals, like lime, generate heat when mixed with water. Various heating pads, operating in this manner, are on the market. When the lime has obsorbed all the moisture it can hold, no further heat emission occurs. One can visualize an engine working on the heat so liberated. The hydrated fuel can be conveyed out to the open air, where solar heat would dry up the moisture and prepare the lime for reuse.

Various electric effects are also possible. Light falling on certain types of surfaces produces an electric current. Solar batteries, which turn light directly into electricity, are now commercially available.

Fig. 176. A 4-inch hole was melted in an iron I-beam placed at the focus of the solar furnace. (U.S. Army photograph.)

Thus far, only small units have been employed, chiefly as a power supply for miniature radios, rural telephone lines, and, most recently, for the radios and other electrical equipment in the earth satellites.

Heating of homes, at least partially, by solar radiation is much nearer to widespread practical realization. The house requires special design, with a large roof area sloping toward the south (for northern latitudes). Glass plates at the top form a narrow green-houselike enclosure, to entrap the sunlight. Within the space one circulates either air or water, the latter in pipes. This heated material is transported to a reservoir for storage until needed. Some houses of this type are already in existence. The fuel necessary to maintain an even temperature has been reduced to less than 50 percent of that for the home without solar heating. Commercialization of various patents is even now under way.

The one great natural collector of solar radiation is vegetation. Chlorophyll, the green of the leaf, has a remarkable chemical action. It causes the carbon dioxide present in the atmosphere to unite with water. The eventual products are cellulose and oxygen.

Fig. 177.   Solar House IV, Massachusetts Institute of Technology Solar Energy Project. Sunlight heats water flowing through a network of copper pipes beneath the sloping glass roof. In the basement the water, pumped from a large storage tank as needed, gives up its heat to air that circulates through the house.

The latter escapes into the atmosphere. The former substance holds, imprisoned in the molecule, the sunlight that was necessary for the reaction to occur.

The process is notoriously inefficient. In the best possible circumstances, vegetation utilizes only a small fraction of all sunlight falling upon a completely forested area. At a 2-percent rate, we should require a minimum of a million square miles of forested land to furnish energy equivalent to our annual consumption of coal.

If I may attempt to forecast the evolution of the power and fuel industries, in the United States, when coal and oil have run out, I should say that large atomic power plants will supply much of the electricity to the larger cities far from hydroelectric developments. Many industries will have moved westward to Colorado, Arizona, and the Pacific Coast, where water power is most plentiful. Wind power will be used in the now desert regions of the Southwest to pump water for irrigation purposes. These arid lands will blossom, as has our Imperial Valley, once water is brought to the surface. The greatest reforestation process in history will cover the entire area with vegetation.

Portable mills will cut the grown timber, systematically and scientifically. Logs are uneconomical to ship. Strategically located giant presses, aided by chemistry, will complete in a matter of seconds the process that took nature millions of years; conversion of cellulose into the ultramodern equivalent of coal—a cubiform, dustless briquette of standard size. There will be no waste. Branches and leaves, as well as the logs, will go through the conversion process. The primary by-product, a syrupy sap, will undergo fermentation, to supply alcohol or perhaps even a supergasoline for the motors of that era. The refuse will furnish innumerable chemical products, to be turned into fibers for clothing, insulation for homes, fertilizers—yes, even food.

From the conversion areas, these products will go forth to the world at large. For the most part, combustion will occur in huge power plants in the suburbs of cities. The consumer will get his fuel as he now does his light, through electric wiring.

If this picture seems fanciful or irrelevant, because of the long-range viewpoint, you may be interested to know that one far-seeing individual has thought otherwise. Godfrey L. Cabot, of Boston, has established two research funds, one at Massachusetts Institute of Technology and the other at Harvard University, to start researches into the problems outlined above. The former is devoted to the development of solar power, the latter to the breeding of new types of trees or plants best suited to the utilization of sunlight.

14

The Sun and the Earth

Not all of our "practical" solar problems are even a few decades distant in time. Some are of immediate interest. I have referred to the so-called *magnetic storms*, which are most violent at times of great solar activity. We recognize such a storm by the presence of fluctuations of the compass needle. Many of you have seen the brilliant northern lights, the aurora borealis, which often occur at the same time. And many types of radio communication—particularly the short-wave, long-distance variety—become impossible during a severe magnetic storm.

Magnetic storms cause effects on the earth's surface as well as in the higher atmosphere. Changes in magnetism induce large electric currents in land lines. The stray currents may blow fuses, cause teletypes to emit unintelligible messages without benefit of operator, or interfere with telephone and power operation. At times of severe

disturbances, large numbers of men must go out to repair the damage occasioned by solar activity.

Explosions on the sun, solar flares and other types of activity shooting out clouds of ionized gas, are responsible for these phenomena. If we could forecast when the sun would erupt, when magnetic storms would occur, operators of short-wave radio stations could plan their schedules of transmission more effectively. This is just one reason why study of the sun is becoming more and more important for economic as well as purely scientific reasons.

⊙

Radio and the Ionosphere

Radio waves are capable of traveling long distances only because the earth's atmosphere contains several layers of electricity at heights of from 70 to 250 miles. The electrification is largely due to ultraviolet energy from the sun, which tears the electrons away from atoms and molecules in the upper air. The ionized layers, which we call the *ionosphere,* act as reflectors for the longer radio waves, so that a signal bounces back and forth between earth and sky, far beyond the visible horizon.

The normal ionosphere consists of at least three distinct layers, called the *E, F1,* and *F2,* whose average heights are about 70, 125, and 150 miles above the surface. We often speak of the lower fringes of the *E* layer, at heights of 50 miles or so, as the *D* region. Here, the radio waves suffer their greatest attenuation by absorption.

The reflections are not exactly mirrorlike. I prefer to think of the ionosphere as a sieve, whose meshes screen the radio waves. The long waves (low frequencies), unable to pass through the mesh, return to earth. The shorter ones speed on through, to be lost in the depths of interplanetary space.

The size of the mesh is continually changing, from a number of causes. The rising and setting of the sun produce fluctuations. Greatest electrification, which corresponds to the smallest mesh, comes around noon, when the sun is highest in the sky. Minimum ionization (largest mesh) occurs somewhat before dawn. Also, there are annual effects, arising from the seasonal orientation of the earth. Solar activity produces additional changes. The ionization tends to be greatest at times near sunspot maximum. Finally, ionospheric storms seem to be associated with magnetic storms, which destroy the normal stratified layers of the upper atmosphere.

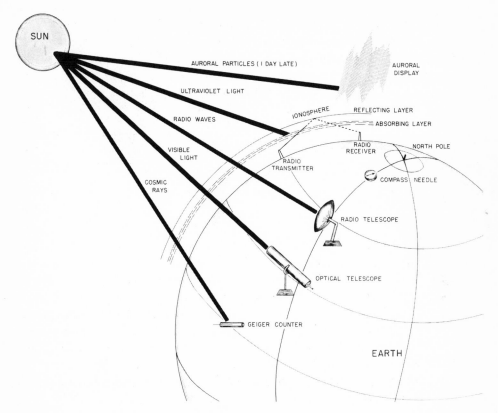

Fig. 178. Representation of solar emission and terrestrial effects during an intense flare. (Covington, National Research Council, Canada.)

All of these considerations are extremely important for radio communications. For successful contact between two radio stations, one must employ a frequency low enough, that is, a wavelength sufficiently long that it will not leak through the ionospheric mesh. Yet, if we take the frequency too low, we also run into difficulty because the longer wavelengths suffer severe absorption in the lower portions of the ionosphere.

Stetson, concerning himself particularly with problems of ionospheric absorption, was the first astronomer to employ radio as a direct tool for study of the sun. His observations of signal strengths of distant radio stations display ups and downs that accord with the sunspot cycle. The studies also show the effects of ionospheric changes of a short-term character. Stetson's work in solar-terrestrial relations in general opened interesting new territory. Many of the

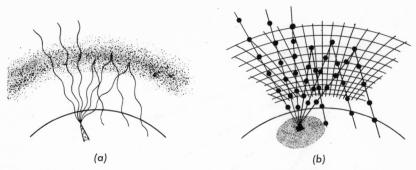

Fig. 179.   Reflection of waves at ionosphere: (*a*) schematic diagram; (*b*) analogy. (Menzel, *Elementary Manual of Radio Propagation*, Prentice-Hall, Inc.)

points now taken as a matter of course were first proposed and investigated by him.

To obtain a worldwide picture, the various governments cooperate in maintaining a network of stations for observing conditions in the ionosphere. Each station consists of a transmitter that sends sharp pulses directly upward and a receiver that records the echo returned from the ionosphere. The time taken for the echo to go up and back measures the height of the radio roof. The mere presence of an echo indicates that the wavelength of the signal exceeds the size of the mesh.

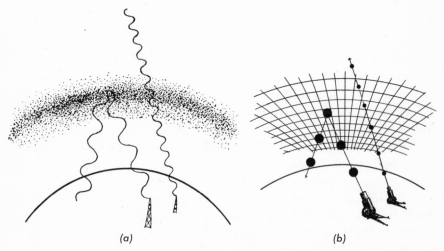

Fig. 180.   Escape of high-frequency waves through ionosphere: (*a*) schematic diagram; (*b*) analogy. (Menzel, *Elementary Manual of Radio Propagation*, Prentice-Hall, Inc.)

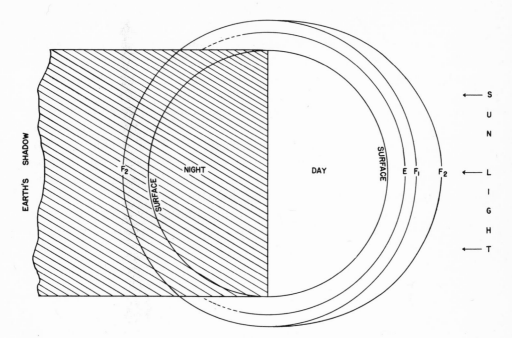

Fig. 181.   Schematic drawing of the three ionospheric layers. (Menzel, *Elementary Manual of Radio Propagation*, Prentice-Hall, Inc.)

When these pulse-echo devices are operated, the basic frequency of the radio waves is slowly changed. Suddenly, at some critical value, the wave sifts through the mesh and escapes into space; at this instant, the echo disappears. From such observations we map, for each hour of the day and for each day of the year, the condi-

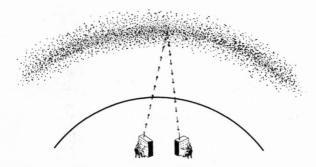

Fig. 182.   Measurement of the ionospheric mesh. (Menzel, *Elementary Manual of Radio Propagation*, Prentice-Hall, Inc.)

tions in the ionosphere in many parts of the world. These maps disclose, on the average, certain regularities of form which enable us to forecast, some months in advance, the state of the ionosphere at any given moment. We might call our prediction a "radio-weather map."

The forecasts of world ionospheric properties, issued monthly by the National Bureau of Standards, provide the basic data for any study of average transmission conditions. They enable the radio operator to choose the frequency most likely to reach the intended destination. The one great problem is the prediction of variations that occur as the result of solar disturbances.

In view of the fact that the entire ionosphere is a phenomenon clearly caused by the ionizing power of solar radiation, we are not

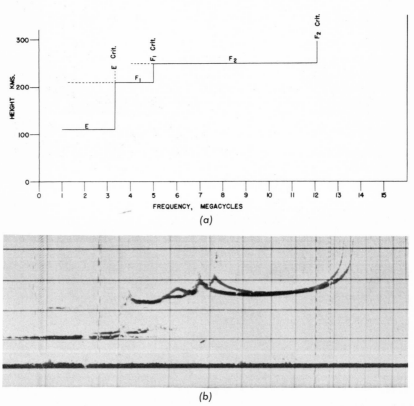

(a)

(b)

Fig. 183.   Ionosphere records: (a) schematic (Menzel, *Elementary Manual of Radio Propagation,* Prentice-Hall, Inc.); (b) actual (Central Radio Propagation Laboratory, National Bureau of Standards).

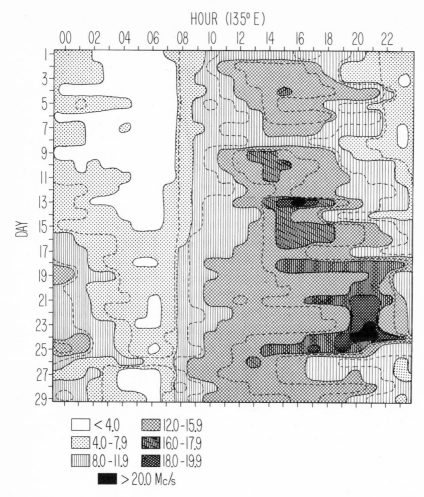

Fig. 184.   Observed hourly values of ionization in the *F2* layer at Okinawa, February 1956. (National Bureau of Standards.)

surprised to find variations attributed to activity of the sun. As an example, Fig. 184 shows the *F2* layer observed at Okinawa during February 1956. The degree of shading indicates the amount of ionization: the black areas show times when there is a high critical frequency, produced by high ionization energy from the sun; while the white areas denote times of lowest ionization. This figure strikingly illustrates the day-to-day and hour-to-hour changes in the condition of the ionosphere. If the sun remained constant, such a diagram would show vertical streaks, the ionization changing only with the altitude of the sun. Everything would be the same for all days of the month, except for the slight progressive shift (or a slant in the vertical streaks) resulting from the changing declination of the sun. Actually, you can see marked fluctuations that reflect the varying activity on the sun. The periods of high ionization have been attributed chiefly to a particular region that crossed the central meridian of the sun on 17 February. Near the west limb on the 23rd, this region produced a famous flare, which was followed minutes later by a remarkable and unusual increase in cosmic-rays observed at many stations throughout the world, and by a geomagnetic storm on the 25th.

We recognize a number of types of sun-caused disturbances in the ionosphere. The first, and perhaps the simplest to explain, is the *radio fadeout,* or *sudden ionospheric disturbance,* abbreviated S.I.D. Dellinger was the first to point out that S.I.D.'s occur at times of solar flares (see Chapters 7 and 8), or brilliant hydrogen eruptions. These flares are commonly found in the neighborhood of sunspots. The great increase of ultraviolet light within such a flare augments the electrification and absorption in the *D* region. A radio wave, previously strong, may fade out and disappear in a matter of seconds. Frequencies in the range of 1.5 to 30 megacycles per second (wavelengths of 200 to 10 meters) are affected, with the lower frequencies most severely absorbed. As one might expect from the nature of the disturbance, only those stations lying in the sunlit hemisphere suffer from S.I.D.'s. The actual fadeout may last from a few minutes up to several hours. Several may occur in rapid succession.

So-called *ionospheric storms* are much more complicated. They begin simultaneously over the entire earth. The layers are often violently disturbed as great clouds of gas, shot from the sun, strike the earth. Sometimes the layers completely disappear, to re-form again hours later. Generally, the ionospheric mesh opens up during the storm, while at the same time absorption increases. The former effect forces us to use lower frequencies, the latter demands the higher ones, so as to minimize the absorption. Often we cannot find a satisfactory compromise between the two conflicting requirements and communication becomes impossible. Since the storm may last for several days, the prediction of communication interruption is a problem of great economic importance.

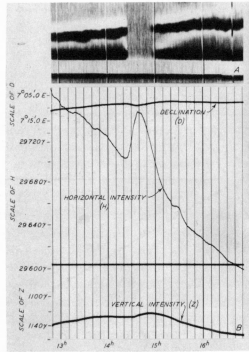

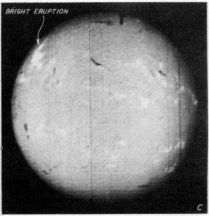

Fig. 185. Changes in the earth's magnetism and cessation of radio reflections from the ionosphere: (*a*) the approximately horizontal line with the bright borders records a signal reflected from the ionosphere; note the interruption of the signal that occurs during the time of the bright eruption (Huancayo Magnetic Observatory, Carnegie Institution of Washington); (*b*) record of the earth's magnetic condition during the same interval; notice especially the disturbance of the horizontal magnetic line during the fadeout (Huancayo Magnetic Observatory, Carnegie Institute of Washington): (*c*) spectroheliogram in hydrogen alpha, 14ʰ42ᵐ (Mount Wilson Observatory).

We do not know the exact cause of such disturbances. The theory most widely accepted is one proposed by Chapman. He suggested that the sun, in an active mood, expels a great stream of ions and electrons like water from a fire hose. The earth, in the course of its motion through space, encounters the stream, which contains positive ions and negative electrons in almost equal number. The earth is a giant magnet, and its magnetic field behaves like a fender, pushing the charges away as the earth plows through the stream. Some of the material forms a ring of ions at high levels, circling the earth's equator. Electric currents in the ring set up a magnetic field opposing that of the earth and tending to reduce the earth's magnetic force. Thus, in broad terms, Chapman accounts for the general features of a magnetic storm. The same stream of ions is presumed to cause the ionospheric disturbances and the aurora borealis, and I shall discuss more recent theories in the section on the aurora.

We have previously noted that the sun is a source of intense radio emission. There are a number of so-called *radio stars* in the sky, each a source of radio waves. It so happens that one of the stronger sources lies in the constellation Taurus, close to the ecliptic. When the sun comes within a few degrees of the Taurus source, we note a fading of its intensity. The radio noise wavers, a sort of radio twinkling which we ascribe to the passage of clouds of ionized gases between us and the sources. The lower frequencies are much more affected than the higher ones, in accordance with theoretical expectation.

I should point out that ionospheric disturbances do not affect all kinds of waves alike. They improve the transmission of the very low frequencies, around 50 kilocycles per second. The increase of signal strength for such frequencies is even more apparent at the time of the S.I.D.'s, which cause the higher frequencies to fade out completely.

The solar radio noise, emitted on frequencies of 30 to 1000 megacycles per second, constitutes a hazard to reception on these frequencies, however. This noise is variable. During World War II, it sometimes interfered with radar operations. With the growth of television and other communication services, the study of causes and effects of solar radio noise becomes ever more a necessity.

⊙

The Aurora Polaris

[This section has been freely adapted from D. H. Menzel, *Flying Saucers* (Harvard University Press, Cambridge, 1953), Chap. 17.]

Few apparitions in the sky are more beautiful or have caused more wonder than the aurora polaris. As the name implies, the "polar lights" are most common in the high latitudes, although occasional displays do appear in temperate regions. The northern lights, or aurora borealis, by virtue of the greater amount of inhabited land in this hemisphere, are widely known. But their southern counterpart, or aurora australis, likewise exists and has furnished a natural illumination to Antarctic expeditions.

Although the farther one goes from polar regions, the less frequent the aurorae become, intensive observation has established

Fig. 186.   The aurora borealis. (Carl W. Gartlein, courtesy of National Geographic Society.)

that the maximum of auroral intensity does not occur exactly at either pole of rotation.

Experiments prove that the earth is a magnetized sphere. It is this magnetism that makes the compass needle "seek" the north and thus guide the traveler. But the magnetic poles do not coincide with the geographic ones. Instead, they lie roughly 12° from the poles of rotation. The earth's north magnetic pole, which we should really refer to as a "south-seeking" pole because it attracts the "north-seeking" end of a compass needle, lies near Baffin Island, north of Hudson Bay, on the continent of North America. The south magnetic pole lies in the Antarctic continent. These poles, moreover, are not precisely fixed. They wander slowly and with some regularity in a roughly circular path around the poles of rotation.

Scientists have found that the aurora borealis occurs most frequently on a circle about 23° from the magnetic pole. From this zone of the auroral maximum, the numbers of visible aurorae decrease both toward the magnetic pole and away from it, toward the equator. At the present time, we on the continent of North America are much more favorably located for observing the aurora borealis than the people at equal latitude in Central Europe, for instance.

Although the forms and patterns of the auroral lights vary as widely as those of summer clouds, we nevertheless can recognize certain definite types. The aurorae are usually classified according to the presence or absence of appreciable ray structure. One of the most common forms is a circular belt of light arching across the northern sky. Some of these arcs merely glow and show no marked internal features, whereas others display a series of rays, like the teeth of a comb. Usually the patterns of light shift slowly, but occasionally the brightness flickers and flashes like rays from a burning forest fire. Thus we get three important classes of auroral forms: the homogeneous arcs, the ray arcs, and the pulsating arcs. When the arcs are so far away that their brightest portions lie below the horizon, only a faint glow indicates the presence of an aurora. Occasionally the rays themselves appear individually or in bundles, sometimes steady, at other times flickering. Not infrequently the auroral light resembles a drapery, which hangs in graceful sweeping folds that may move back and forth like a curtain or a long skirt swaying in the breeze.

When the arcs are not well defined, the northern sky may be traversed by homogeneous or rayed bands. Occasionally we see only a uniform, diffuse, pulsating surface.

In very intense displays, the auroral glow will sometimes reach to the zenith or beyond and form the beautiful crown or corona. The corona usually appears as a series of streaks radiating from a dark center, which lies in the direction toward which a compass needle would point if it were free to move vertically as well as horizontally. We call this point the *magnetic zenith*.

Many advertising signs, like the red neon lights, shine because electrons, driven through the gas, smash into the neon atoms with sufficient force to make them radiate. The auroral glow arises when protons and electrons from the sun smash into various atoms and molecules that occur in the earth's upper atmosphere. The characteristic green glow comes mainly from oxygen. Under certain circumstances this atom can also emit a reddish light. Nitrogen, the most abundant constituent of the earth's atmosphere, can also produce a deep red glow.

Flares and other types of solar activity eject the protons (hydrogen nuclei) and electrons responsible for the northern lights. For a long time all evidence for a solar cause was indirect. But in the past few years, Meinel, Gartlein, and others have discovered that auroral displays sometimes show radiations from the very light atom, hydrogen. Their spectroscopic observations—of Doppler shifts—indicate that the hydrogen gas, instead of being stationary in the earth's atmosphere like the nitrogen and oxygen, is rushing toward us with a speed of from 200 to perhaps 2000 or 3000 miles per second. These observations provide direct evidence of clouds of gas coming to the earth from the sun.

Major disturbances of the earth's magnetic field often accompany intense auroral displays. A compass needle points toward the north magnetic pole. If our compass is very sensitive, however, we soon discover that the needle is never really still. It drifts first eastward and then westward, making a fairly regular progression in the course of 24 hours. The normal movement is small. The tip of a hundred-foot compass needle would wander less than an inch during a day. But the total amount of the drift and the smaller fluctuations associated with it change markedly, sometimes from just one day to the next. Days when the needle shows a big fluctuation we call *magnetically disturbed.*

Fig. 187.   Great solar flare of 25 July 1946, hydrogen record: (*left*) 15^{h}37^m U. T.; (*center*) 16^{h}14^m; (*right*) 16^{h}24^m. Note also Fig. 188. (d'Azambuja, Meudon.)

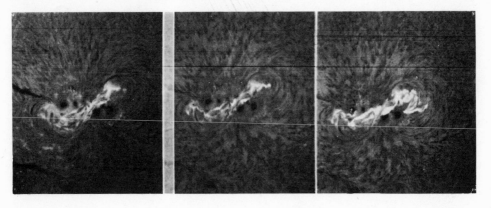

Fig. 188.   Great solar flare of 25 July 1946, hydrogen record, 17ʰ30ᵐ U. T. Note also Fig. 187. The flare exhibits an unusual spiral symmetry with respect to the bipolar spot group. (d'Azambuja, Meudon.)

The distribution of the aurora borealis on the surface of the earth clearly suggests that magnetism has something very definite to do with the phenomenon. Theories of the aurora borealis based on solar and terrestrial magnetism go back to the early 1900's, when the Norwegian scientists, Birkeland and Störmer, were carrying out basic studies of the aurora and its behavoir, in the laboratory, in the field, and by mathematical calculation.

Störmer first worked out in detail how an electrically charged particle, say an electron, might traverse the distance from sun to earth and get caught in the earth's magnetic field in such a way as to produce the aurora. He showed how electrons would tend to spiral around the earth's magnetic lines of force and how magnetism could guide the electron to earth.

It is particularly important to emphasize that the electron speeding in its path from sun to earth, on any theory of magnetic control, derives no motive power at all from the magnetic field. The electron must be able to come all of the way on the initial shove it gets from the sun. The magnetic field is just the "track" about which the electrons spiral. Where the field is weak, the spiral is large; where the field is strong, the spiral is much tighter. The faster the particle moves, the fewer loops it will take in its path from sun to earth.

Low-energy particles shot from the sun would move along tracks much like that of Fig. 189. The charged particles would never reach the earth's equator. In fact, they would be concentrated in a

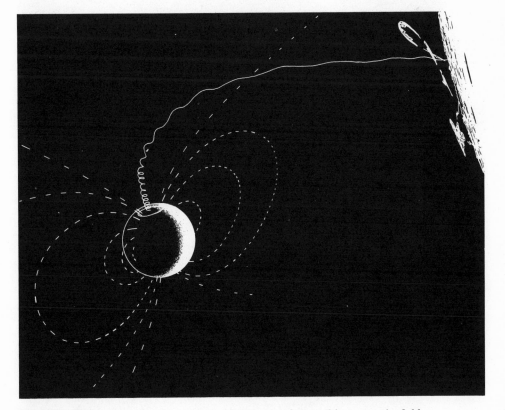

Fig. 189. The path of a low-energy electron in the earth's magnetic field, according to Störmer.

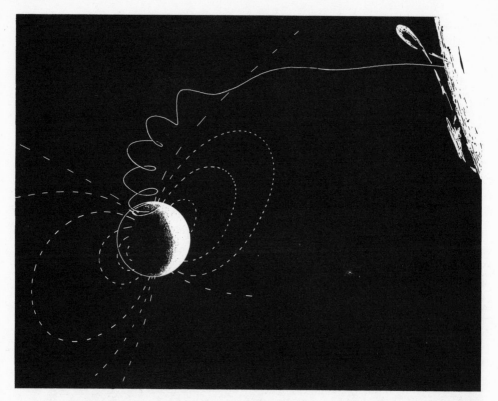

Fig. 190.   The path of a high-energy electron in the earth's magnetic field, according to Störmer.

very tiny ring around the earth's magnetic pole. Störmer quickly saw that his elementary theory needed elaboration to explain the auroral zone. Thus he postulated greater energy for the electrons so that, instead of falling close to the pole, they would settle down along a circle about 23° from the pole—the auroral zone—as shown in Fig. 190.

Sydney Chapman showed that Störmer's idea of a sun shooting large quantities of electrons to the earth was inconsistent with other facts. Each electron, for example, carries with it a negative charge when it leaves the sun. Thus the more electrons that have been ejected from the sun, the greater the positive charge that the sun would have. It is well known that opposite electric charges attract one another. Very few electrons would leave the sun before the positive charge on the solar surface would be great enough to prevent any further escape of the electrons. Indeed, the total number of electrons that could escape from the sun would be able to run a one-cell flashlight for less than a minute. Thus, this theory could not possibly account for the enormous energies and luminosity of the aurora polaris.

Chapman and his colleague Ferraro theorized about the clouds of gas escaping from the sun. They reasoned that, since gases were escaping, each negative electron must be accompanied by the positively charged remainder of the atom from which it was originally torn. An atom that has lost an electron we call an ion, and these great clouds of ionized gas, in Chapman's theory, replaced the stream of electrons that had appeared in Störmer's calculation.

Chapman showed that if such a cloud of particles were shot out from the sun, like water from a slowly rotating fireboat, we should have a situation that develops something like that shown in Fig. 191. The sun rotates once about every 27 days. Since the equatorial regions move a little more rapidly than those of intermediate latitude, no single figure can represent the time of the sun's rotation. However, the sun does spin around more rapidly than the earth moves forward in its orbit. Hence, any cloud of gas shot from the sun would curve as shown in the diagram and strike the earth from behind.

At considerable distances from the sun, as Chapman and Ferraro showed, and also as D. F. Martyn of Australia later indicated in greater detail, the clouds would move almost unhindered. The magnetic fields in space are so weak that they could not possibly serve as a guide rail, as Störmer hypothesized. As a matter of fact, the moving gas would tear through any magnetic field present, and, to continue with the track analogy, rip up the rails and sweep through space almost unhindered by the magnetic lines of force. More accurately, the rails will stretch, but never will an open track occur. About the only effect that the magnetic field could have would be a minor focusing action, which would tend to keep the cloud from dispersing.

Such a cloud of ionized gas could not penetrate all the way to the earth. At a distance of three or four diameters from the surface the cloud divides, leaving the earth in a hollow. It is the earth's magnetic field that produces this effect, raising a sort of cosmic "umbrella" to fend off the rain of ionized gas from the sun. Thus, if conditions were absolutely perfect, none of the material would

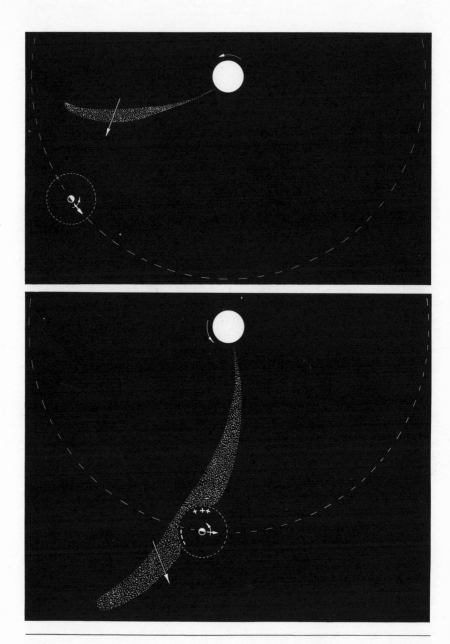

ever penetrate to the surface of the earth, and we should never
have an aurora borealis.

I have made calculations based on a new theory called magneto-
hydrodynamics. These studies indicate that the simple conditions

Fig. 191.   Three stages, according to Chapman, in the envelopment of the earth by a surge of gas from the sun. Note that the earth's magnetic field forms a protective envelope.

exist only when the cloud of gas is absolutely uniform, without any appreciable clumps of gas within it. If such a cloud possesses a sharp irregular edge, it will put a severe strain on the earth's magnetic umbrella. The magnetic lines of force are flexible and give. Like an umbrella in a howling gale, the magnetic field can sometimes turn inside out, and if this happens, the umbrella no longer can protect the earth from the storm of solar gas. An umbrella that is blown inside out becomes a "funnel" as shown in Fig. 192a, b.

The weakest portion of the umbrella lies on the afternoon and evening side of the earth, since the ion clouds tend to overtake the earth from that direction. But the greatest caving occurs in the polar regions where the lines of force—the ribs of the umbrella— extend farthest. Thus the earth's magnetic umbrella is almost completely gone within the auroral zone. And when, during a phase of

(a)

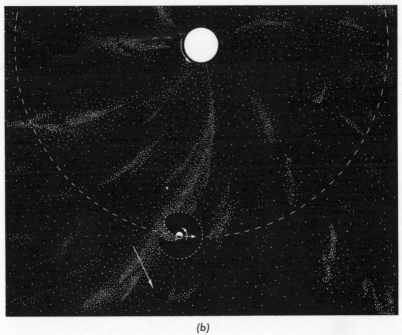

(b)

(c)

Fig. 192.   Schematic formation of aurora, according to Menzel. Clouds of gas encounter the earth and bend its magnetic field inward.

particularly great solar activity, any exceptionally heavy or dense cloud of gas reaches us, the magnetic lines of force can be damaged to a lower latitude, the funnel opened more widely, and more material sucked in by what is a sort of gigantic vacuum cleaner (Fig. 192c).

The end of the funnel near the earth has a long narrow opening, roughly parallel to a line of magnetic latitude. Thus the material shooting down through the funnel is concentrated and tends to appear as a curtain, or, if the distribution is irregular, as a series of pulsing rays.

Observations from artificial satellites have disclosed the presence of two belts of highly energetic radiation—ions and electrons—apparently entrapped in the earth's magnetic field. These so-called *Van Allen belts,* Gold has shown, may also play a role in the production of aurorae. For, as ejections from the sun pommel the earth, distorting the outer magnetic field, the inner field also changes form. The belts of entrapped particles alter their positions, sometimes acquiring energy during the process, enough energy to penetrate the earth's atmosphere and produce auroral displays.

The amount of radiation in at least the outer of these Van Allen belts appears to fluctuate considerably. The solar corpuscular radiation appears to exert a heating effect on the earth's outer atmosphere, causing it to expand upward. Jacchia has shown that artificial satellites moving through the ionospheric regions encounter greater atmospheric resistance during times of maximum solar activity.

In August-September 1958, U.S. scientists performed, under the secret code name *Argus,* what some persons have termed the most remarkable experiment in history. They exploded three hydrogen bombs at an altitude of about 300 miles above the surface of the earth, within the upper ionosphere. The explosions injected ionized hydrogen into the region, which behaved like the gas in the Van Allen zones injected by solar activity and trapped in the earth's magnetic field. Intense auroral displays occurred. Indeed, some scientists, themselves not in on the secret, inferred the occurrence of an atomic blast by virtue of lithium lines seen in the spectra of these aurorae. (Lithium, itself a fusible element, has generally been recognized as an ingredient of the hydrogen bomb.)

The beauty of the aurora polaris, with its wide variety of forms and its ever-changing appearance, is something that everyone should see. But when the auroral light glows in the sky, we may recall the superstitious horror that such apparitions have occasioned throughout the ages. We can enjoy and appreciate its beauty, because we know and understand what forces produce it and realize that it is not a supernatural apparition or sign in the sky. As yet, we cannot predict it as accurately as we can solar eclipses, which have also caused their share of fear and horror. Nevertheless, we are making progress in our understanding of the sun and of how solar and terrestrial magnetic fields control the motions of gases in their flight from the sun to earth.

⊙

Prediction of Magnetic Storms

We know that magnetic and ionospheric disturbances and bright aurorae are more violent and more frequent around sunspot maximum than around sunspot minimum. The correlation with individual sunspots, however, is not striking. The majority of the very greatest disturbances can be attributed to a large sunspot around the solar central meridian at the time, often to a very large flare occurring 1-2 days previously. But the inverse relation is less satisfactory. Many other sunspots, of comparable size and activity, move across the disk of the sun without disturbing the earth; large flares also occur without terrestrial consequences.

Since the correlation does not work in both directions, we are forced to infer that sunspots as such are not fundamental. Recent research has accordingly been directed toward study of the various properties of sunspots, in the attempt to discover some criteria that will enable us to distinguish storm-producing sunspots from the others prior to the onset of a storm. This approach has produced several clues but as yet no complete solution. For example, Newton has shown that spots producing an above-average number of flares are more likely to disturb the earth than those subaverage in flare production. Denisse and co-workers find that spots associated with an enhanced level of radio noise radiation from the lower corona at 168 megacycles per second tend to be accompanied by terrestrial disturbance; whereas spots without any enhancement of the radio noise tend to be associated with unusually quiet geomagnetic conditions. Bell and Glazer find that spots with complex (γ) magnetic fields (see p. 117) are more likely to disturb the earth than are simple bipolar (β) and unipolar (α) spots.

These correlations appear to be in harmony with recent advances in our knowledge of sunspots, which indicate that magnetic fields tend to inhibit rather than enhance the normal convection of the solar atmosphere (p. 127). On the peripheries of large spots, we expect increased convection, to compensate for the lower radiation over the darker spot area. The complex magnetic fields of the more active spots may well be the result rather than the cause of such convective flow. The turbulence may churn up the magnetic lines of force and the associated electric currents, redistributing them in a complex, irregular pattern. But much remains to be done before we can expect to make satisfactorily accurate predictions of the magnetic and ionospheric disturbances.

One persistent effect, which certainly points toward the existence of narrow corpuscular streams from sun to earth, is the tendency of storms to recur at fairly regular intervals of approximately 27 days. This figure corresponds closely to the mean rotation period of the sunspot zones. If the sun is like a slowly rotating fireboat with hoses squirting in several directions, we should expect to get a splash of corpuscles each time the hose turns in our direction.

This tendency of storms to recur at 27-day intervals is more conspicuous among storms of moderate intensity than among the very great storms, and becomes more pronounced on the declining branch of the sunspot cycle than when spot numbers are high. The recurrent storms of moderate intensity, unlike the great storms discussed above, can only rarely be associated with a prominent sunspot. On the contrary, as Allen has shown, these recurrent storms tend to avoid large sunspots. They sometimes happen when the solar disk is altogether devoid of sunspots. Bartels named the mysterious areas of the solar surface responsible for these magnetic disturbances *M-regions*. The identification of the M-regions with some definite solar formation has long been a problem.

Ordinary prominences seem to cause no marked terrestrial disturbance. Attempts have been made to identify the M-regions with bright coronal patches. Some association exists between the green coronal line, the patches of faculae, and the general pattern of spotted areas. The behavior of the ionosphere during total eclipse, as the moon blots out one bright coronal region after another, seems to indicate that these regions, rather than the entire solar surface, are responsible for the ordinary ionizing radiation, as Pierce and Waldmeier have shown independently. The change in the ionospheric condition is not continuous during an eclipse. The records exhibit abrupt discontinuities each time the moon covers a bright coronal area.

Recent studies, however, point more in the direction of a negative relation between bright coronal areas and terrestrial disturbances of the recurrent and moderate type. Since the eclipse effects arise from electromagnetic radiation, whereas the storms result from corpuscular beams, the two consequences are not necessarily contradictory. From observations in 1942–1944 and 1950–1953, on the declining part of two sunspot cycles, several workers have found that conditions of unusual quiet, rather than storms, tended to occur after a bright coronal region crossed the central meridian of the sun. In these same periods, Bell and Glazer, and Bruzek, have found evidence that M-regions may be areas of unusual weakness in the green coronal line. They find that magnetic storms tend to occur 1–4 days after central meridian passage of regions of unusually faint green-line corona. Since the brightness of the green coronal line is related to the density of the coronal gases, Bell and Glazer suggest that M-regions may be areas of low coronal density, and that an inverse relation exists between the density of gas in the corona facing the earth and the density of corpuscles in the neighborhood of the earth. The Babcocks suggest that areas of the solar surface characterized by weak unipolar magnetic fields may be the M-regions, and that this type of field makes them able to squirt out the streams of corpuscles.

Unfortunately the relations of the coronal line to geomagnetic and auroral conditions which I have described above appear to break down at spot minimum and to apply only in the last 3–4 years of a cycle.

At all stages of the solar cycle, the magnetic and auroral activity and ionospheric disturbances show pronounced maxima in March and September, with minima in June and December, as shown in the Fig. 193. Two explanations have been proposed, which we may call the equinoctial and the axial. The former suggests that the

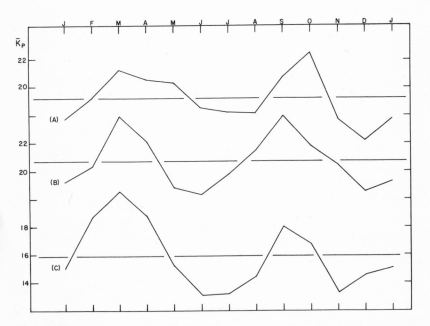

Fig. 193.   Average monthly values of K_p, a measure of world-wide geomagnetic disturbance, for three phases of the sunspot cycle: (A) maximum, (B) decline, and (C) minimum. (Bell and Glazer.)

earth responds more violently to streams of solar corpuscles during the equinoctial periods. The latter derives from the fact that the plane of the earth's orbit is inclined 7.2° to the sun's equator. The heliographic latitude of the earth varies from 7.2° north of the solar equator in early September to 7.2° south in early March (Fig 25). We see a maximum area of the sun's northern hemisphere in September and a maximum area of the southern hemisphere in March, at approximately the times of greatest geomagnetic activity. The magnetic activity is least around the time when the earth is crossing the solar equator. The axial explanation thus suggests that solar corpuscles more readily intercept the earth when the earth is farthest removed from the equatorial plane of the sun. Spots and bright coronal patches occur most frequently in heliographic latitudes 5–25°, while the equatorial belt itself contains few spots. If the corpuscular streams start from the spot belts and leave the solar surface in a radial direction, corpuscles from the northern and southern belts are most likely to intercept the earth in September and March respectively.

Probably both the equinoctial and axial factors contribute. Bell and Glazer obtain their best correlations between geomagnetic conditions and brightness of the coronal line on the same side of the solar equator as the earth. They find negligible correlation with the corona on the opposite side of the equator from the earth. On the other hand, the great magnetic storms show about an equal frequency of association with spots on either side of the solar equator.

The identification of M-regions and the discovery of the best possible criteria for distinguishing the storm-producing sunspots are of great practical importance. For only then can we hope to forecast, with some precision, when the disturbances will occur. Radio companies will then be able to plan their communication schedules and select frequencies with greater assurance that the transmitted messages will arrive safely. Solar research has already contributed significantly to the solution of these radio problems.

$\odot$

Cosmic

Rays   In a previous section I mentioned a remarkable increase in cosmic rays associated with a solar flare on 23 February 1956. Although the most recent was by far the largest, four previous such increases have been observed—the first on 28 February 1942, the others on 3 March 1942, 25 July 1946, and 19 November 1949.

Before I proceed further—What are cosmic rays? Very briefly, they are atomic nuclei stripped of their electrons and moving at very high velocities, up to nearly the velocity of light (186,000 miles per second). All of them have very great energies, even those spoken of relatively as "low-energy" rays. Most of them are protons but scientists have found also a sprinkling of heavier nuclei. Cosmic rays, as their name would imply, impinge upon the earth from unknown sources in outer space. Because they are charged particles, the earth's magnetic field and also magnetic fields in space influence their motions. Only particles of relatively high energies can get in through the earth's "magnetic umbrella" at the equator, whereas particles of lower energy get in at higher latitudes. The relative frequencies of various energies can be determined from a study of the variation of cosmic-ray intensity with latitude.

High-energy cosmic rays have an enormous penetrating power, and will go through many feet of concrete. Even a large building does not provide significant shielding from the most energetic varieties. Though you are unaware of the events, many such cosmic "bullets" shoot through your body every second.

The origin of cosmic rays is one of the most fascinating unsolved problems of modern science. Where in the universe can particles possibly acquire such tremendous energies, to move with nearly the velocity of light? A discussion of the various theories is beyond the scope of this book, but there are a few points to be made in connection with the sun.

The high-energy cosmic rays are definitely not of solar origin and are not influenced by solar activity. Also little or none of the daily supply of lower-energy cosmic rays is believed to originate in the sun. However the five flare-associated events that I mentioned at the start of this section establish that the sun does, at least on rare occasions, emit low-energy cosmic rays. Gold has suggested that other major flares may also emit cosmic rays, which are prevented from reaching the earth in detectable numbers because of unfavorable configurations of magnetic fields in interplanetary space.

In addition, the sun appears to influence the density of cosmic-ray particles indirectly. An event more common than the five increases in 14 years already mentioned is the *Forbush decrease,* an

occasional decline of a few percent in the abundance of low-energy
cosmic rays arriving over a few hours, followed by a slower recovery
to normal. These often, but not always, occur in conjunction with
magnetic storms and appear undoubtedly associated with solar
activity. Also, Simpson and co-workers at the University of Chicago

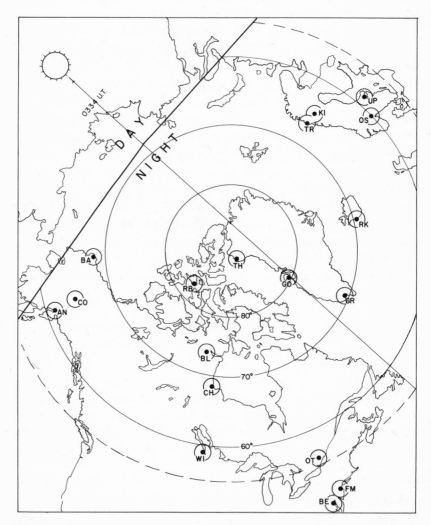

Fig. 194.   Location of night vertical sounding stations, recording ionospheric disturbances associated with flare and cosmic-ray increase of 23 February 1956.
(National Bureau of Standards.)

have observed a 27-day variation of several percent in the intensity of low-energy cosmic rays, a recurrent decrease.

Both of these types of decrease, Morrison, Gold, and others have suggested, possess an indirect solar cause. They suggest that decreases result when a cloud of gas containing magnetic fields is shot out by the sun and expands rapidly in the space near the earth. The expanding magnetic field tends to deflect cosmic rays and thus in some measure to shield the earth.

Two other phenomena may also be explained by the screening effect of expanding magnetized clouds from the sun. Forbush has found that the abundance of low-energy cosmic rays show a slight variation over the sunspot cycle, in the sense of a negative relation to the sunspot number. The amplitude of this effect increases toward the lower energies. Also the minimum cosmic-ray energies observed around sunspot minimum are less than those measured around sunspot maximum. When solar activity is high, the sun ejects numerous clouds of expanding magnetized gas and the cumulative effect reduces the abundance of low-energy cosmic rays by a few percent. The relative absence of such clouds around spot minimum permits less energetic particles to come in.

The ionospheric data from the aforementioned flare of 23 February 1956 are illuminating. Ionospheric absorption is presum-

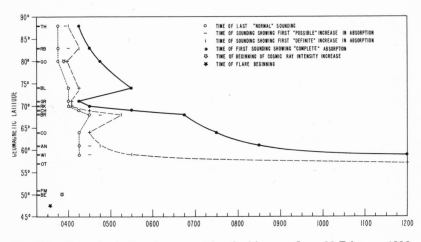

Fig. 195.   Ionospheric disturbance associated with great flare, 23 February 1956; times of onset of intense ionospheric absorption at stations shown in Fig. 194. (National Bureau of Standards.)

ably due to the arrival of ions and electrons from the sun. The fact that they affect stations in the night hemisphere (Fig. 194) indicates that the radiation has suffered deflection, presumably by magnetic fields in space and near the earth. Figure 195, depicting the times of the onset of complete ionospheric absorption, shows that the disturbing clouds arrive sooner at high magnetic latitudes than at lower ones. It also shows the ease with which the clouds penetrate the auroral zone at latitude 70° or so. And Fig. 196 shows that complete ionospheric blackout tends to persist longer at higher geomagnetic latitudes.

⊙

Weather

Forecasting   Solar eruptions affect, in addition to the ionosphere, another layer of atmosphere, a region some 15 to 30 miles above the earth's surface, in which considerable quantities of ozone exist. This is the gas

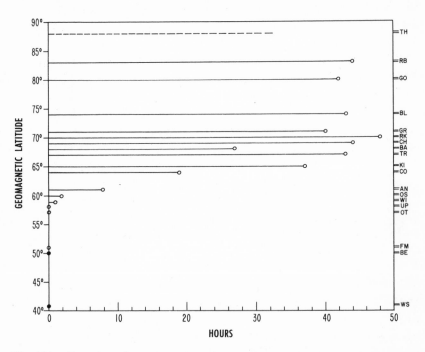

Fig. 196. Duration of complete absorption associated with flare of 23 February 1956; see Figs. 194 and 195. (National Bureau of Standards.)

that screens off sunlight in the far ultraviolet, as I have previously mentioned. Apparently the layer is important in a number of ways. For one, the layer absorbs a sizable amount of ultraviolet energy. This energy tends to warm the layer. Some meteorologists believe that the alternate heating and cooling of this layer is a mechanism by which the sun transmits its variations to the lower levels that control terrestrial weather.

Weather! Astronomers can predict to within a few seconds an eclipse that is to occur a century hence. The meteorologist is uncertain of even tomorrow's weather. Why this disparity of accuracy in the forecasting of two different natural phenomena?

The reason is clear. Only a few celestial bodies—the sun, the moon, the earth, and, to some extent, the other planets—enter into the calculations of eclipses. Gravitation alone is the governing law. But in the case of the weather we are dealing with the myriads of molecules of the earth's atmosphere. Gravitation plays a part, but we also have countless other factors: heating by the sun, the current pattern of cloudiness, the blowing of the winds, the general plan of the earth's circulation, the presence of mountains and plateaus, forces of the earth's rotation. In addition, the future depends appreciably upon past history.

We do not know how solar variation influences the weather. Most meteorologists are fairly well convinced, however, that the sunspot cycle has some effect on weather trends.

Certainly the sun is the primary factor governing weather conditions. The mere existence of the seasons is proof enough. We have winter when the sun's rays strike glancingly and summer when they are more nearly vertical and thus concentrate their heat on a smaller area of surface. Each year the relative geometry of the earth and sun repeat exactly. But weather, from year to year, is by no means constant. We have warm winters and cold ones, wet summers and dry ones. Can such changes be due in any part to solar variability?

Perhaps one of the most convincing proofs of a relation has been found by A. E. Douglass, in Arizona. Douglass has collected cross sections of tree trunks from thousands of different sources, from living trees to those that flourished millions of years ago. The annual rings of the trunks form a natural weather record. A thick ring indicates a favorable growing season, perhaps warm with lots of rain; a thin ring indicates just the opposite. It is interesting to note

Fig. 197.   Tree rings and the sunspot cycle. An Eberswald (German) pine, show-ing large growth near sunspot maxima. (Douglass.)

that all the trees from a given locality show the same pattern of weather. To match the records of a pair of trees we must, of course, allow for the fact that they may have sprouted at different times. In this way, by fitting the series of overlapping records, Douglass has been able to extend our knowledge of weather back thousands of years.

The tree trunks show that the earth has been experiencing weather cycles throughout the centuries covered by the study. Moreover, the pattern seems to display the 11-year sunspot cycle. One example of a tree-ring record appears in Fig. 197.

Many other meteorological studies indicate the general existence of an 11-year or 22-year cycle in weather phenomena. The associated effects include: rise and fall of lake levels, changes in barometric pressure, rainfall. Recent correlations, made by R. Craig, by R. Shapiro, and by W. O. Roberts and co-workers, suggest that solar activity, as measured by geomagnetic disturbance, is closely linked with major changes in the pattern of weather over the earth's surface.

One of the elusive qualities of weather prediction is the nature of the forecasting process itself. The basis is empirical, dependent upon previous weather records. Our meteorological observations show a certain condition, such as distribution of temperatures, air pressures, winds; we search past records for similar circumstances and base our forecast upon what happened then. The method is fairly effective, but it avoids any reasoning as to why the particular weather sequence followed.

Of course, not all of the modern procedures are quite so unconcerned with cause and effect. A new meteorological school has recently arisen, founded on work by Rossby, Willet, Wexler, and others. Movements of different types of air mass—polar air, tropical air, and so forth—are considered fundamental. Haurwitz finds a relation, for example, between variations in the ozone layer and movements of air masses.

In the future, it seems that solar studies will play a growing part in the forecasting of weather. The observations may contribute to an increased accuracy of the short-range forecasting. But perhaps their greatest potential value is the improving of long-range forecasting. If, for example, we could tell the farmer that the next growing season will be hot or cold, wet or dry, late or early, how

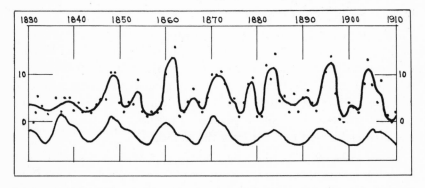

Fig. 198. Tree growth (*upper curve*) and sunspot numbers (*lower curve*), from a study of living trees in England, Norway, Sweden, Germany, and Austria. The tree-growth curve shows strong maxima near the sunspot peaks, with weak maxima roughly halfway between. The two maxima in the 11-year cycle correspond to the general rainfall curves, which show similar periodicity. (Douglass.)

Fig. 199.   W. O. Roberts, Director of the High Altitude Observatory of the University of Colorado.

valuable such information would be! The farmer would then know what to plant and when to plant it. We should have far fewer crop failures.

In some sections of the country, the farmers use rules of thumb that are often superstitious nonsense. As an example, I give the following rule, widely used in the southeast United States. Take the 12 days immediately after Christmas and note the general character of the weather of each: warm, cold, wet, and so on. Associate each of these days with a month of the year and you will find—so the argument runs—that the weather of that day also determines the weather of the associated month. The concept is ridiculous and utterly unscientific. I cite it primarily to show how strong a need the farmer himself feels for long-range forecasts.

Improvements in weather forecasting, like those in the prediction of radio disturbances, have a wide variety of applications. Both are particularly important in the field of air-travel safety. Any increase in the reliability of weather or communication conditions will necessarily benefit the air industry, which is so very dependent, for successful operation, on knowledge of these phenomena.

The studies, incidentally, will give valuable information also concerning conditions in the upper regions of the earth's atmosphere: levels our future rocket ships will traverse.

⊙

Before I bring this discussion to a close, I want to mention a few more possible applications of solar studies. I have referred several times to the screening effect that the ozone layer exerts on ultraviolet radiation. The energy that just manages to filter through has an intense biological effect. One of its specific qualities is the ability to kill bacteria. We know that the amount of ozone varies with the sunspot cycle. Furthermore, we recognize that the amount of this lethal ultraviolet undergoes some variation in consequence. We infer—and I emphasize that the inference is not proved—that the well-known cycles in various epidemic diseases may be traceable, in part, to an effect of solar variability.

Correlations of a much less probable nature have been suggested, although some of the curves look convincing. These relate to periodicity in the abundance of certain types of wild game, the quality of their furs, the multiplying of insect life, and so on. As DeLury has noted, ultraviolet light might conceivably produce such periodic effects. The variations are certainly real and many of them correspond roughly with the sunspot cycle. Even exact agreement, however, cannot prove a direct cause-and-effect relation. One must discover and trace the details of the physical processes, in order to bolster up the evidence of statistics.

As an illustration of the power of statistics, I cite a true example. Some years ago a British scientist decided to study the relation between births and weather. He came to the somewhat surprising conclusion that more babies were born on clear days than on stormy ones, and finally published the results. One of his colleagues, who could see no logical reason for such a relation, decided to check back. Step by step, he traced the mathematics and found no error. Finally, in desperation, he went to the bureau that supplied the original information. Everything was correct—except that the clerk had inadvertently copied, not the date of birth, but the date upon which the parent had registered the birth! The statistics proved merely that people do not like to go out in inclement weather.

Almost all phenomena show cyclic behavior: plants, animals, man and his affairs. One relation frequently discussed is the stock market. The question has often been raised whether the sunspots have some effect on the bullishness or bearishness of the market. The idea is not new. Sir William Herschel himself suggested a possible correlation between sunspottedness—the numbers had not then been invented—and the price of wheat. As an example I give here two curves which greatly extend Herschel's relation (Fig. 200). One is for the price of wheat, adjusted for the gradual rise in price over the long interval, so that only the variations remain. The other is for sunspots. In certain regions the curves correspond surprisingly well; in others, the values oppose one another.

Many organizations are furnishing (at a price) advice to investors on the probable cyclic behavior of the stock market, by comparison with types of natural cycles such as sunspot numbers, planetary aspects, weather records, and so on. Since both the market and

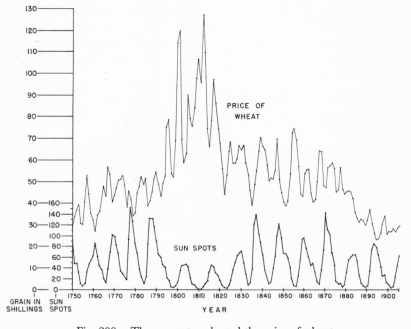

Fig. 200.   The sunspot cycle and the price of wheat.

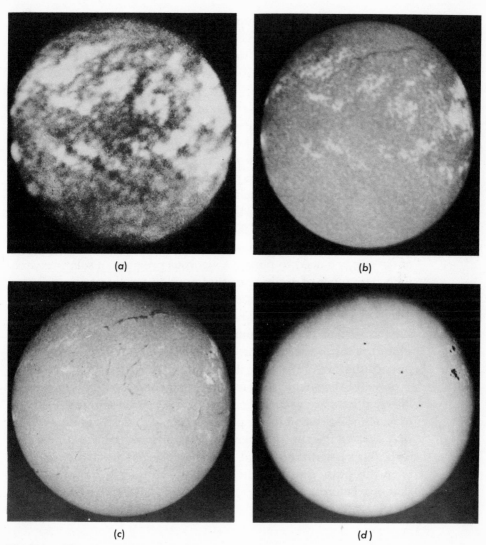

(a)   (b)

(c)   (d)

Fig. 201.   The sun, 13 March 1959. (a) This, the first photograph made of the sun in the ultra-violet light of hydrogen Lyman-alpha, 1215 A, was obtained by J. D. Purcell, D. M. Packer and R. Tousey of the U. S. Naval Research Laboratory. The picture was taken from a rocket at an altitude of 123 miles above White Sands. The instrumental resolving power was about ½ minute of arc. North is at the top, east to the right (Official U. S. Navy Photograph). (b) Comparison Ca K spectroheliogram (McMath-Hulbert Observatory). (c) Ordinary Hα photo-graph (Naval Research Laboratory). (d) White-light image (U. S. Naval Observatory).

natural phenomena have ups and downs, predictions of one based upon the other may achieve some measure of success for a limited time, *if* the curves happen to coincide. But the indications are generally against the reality of any connection. Temporary success is largely a matter of chance.

Most scientists consider the case for such relations as definitely unproved. The stock market is such a complex phenomenon, involving supply and demand, taxation, legislation, and so many other human factors, that I cannot see why a solar relation should be expected. Or, if it were present here, one would suppose that the relations would be much more outstanding in the subordinate factors, such as the abundance of raw materials, which is less affected by purely human factors.

With benefits such as those implied above to come from studies of the sun, we need to pursue a vigorous program of solar research, with new types of equipment that employ the latest scientific improvements, including observations from rockets and satellites, if not from the moon itself. The research program should emphasize such aspects as the development of new indices of solar activity. It should point toward a thorough theoretical understanding of the physical nature of the solar-terrestrial relations.

Some of the suggestions made in this chapter are highly speculative; others are extremely practical. We are dreaming a dream, in which solar research is to play a practical role. In the future, man may look to the sun not merely as the giver of light and heat, nor with the superstitious mind of the astrologer, but with the firm scientific faith that—in the sunlight—coming events cast their shadows before.

Index

Abbot, C. G., 71, 72, 78, 80, 81, 159, 297, 298
Abetti, A., 120, 121
Absorption: ionospheric, 305; terrestrial, 70–74
Absorption lines, 85
Absorption spectrum, 56–57
Adams, W. S., 103
Age: of Earth, 2, 258, 278; of universe, 280
Alchemy, modern, 263 ff
Alfvén, H., 126, 220
Allen, C. W., 202, 326
Aller, L. H., 86, 92, 220
Alpha particle, 259, 260, 261, 262
American Association of Variable Star Observers, 26, 129, 157
Anaxagoras, 9, 10; geometry of, 10
Angstrom, A. J., 48
Angstrom unit, 48
Antares, 274
Antimatter, 261
Antiproton, 260, 261
Argus experiment, 324
Aristarchus, 11, 12
Aristotle, 10, 11, 30
Athay, R. G., 90, 168
Atmosphere: earth, 70–74, 122–124; opacity, 158–160; planetary, 94–95; solar, 27, 86, 88 ff, 121, 127, 158–160, 161
Atom: ancient concept, 52; chemical behavior, 52–53, 57–58, 59; characteristic wavelength, 46, 47; empty space in, 55; ionized, 59; orbit models, 54–55; radiation by, 46–47; radio-

active, 1, 4, 259; structure of, 52 ff; velocities, 83; wave model, 56–58. *See also* Elements, Nucleus
Atom smasher, 264
Atomic bombs, 267, 268
Atomic collisions, 84
Atomic energy. *See* Energy
Atomic nucleus, 55, 259 ff. *See also* Nucleus
Atomic power, 294, 301
Atomic reactor, 295
Atomic structure, 52 ff
Aurora polaris, 303, 311, 312–324; appearance, 312, 313; Argus experiment, 324; cause, 314–315; distribution, 313–314, 316; magnetic storms, 315; magnetic zenith, 314; spectrum, 315; sunspot cycle, 324; theories, 316–324; Van Allen belts, 323, 324; zone, 313–314, 321

Babcock, Harold, 114, 115, 327
Babcock, Horace, 114, 115, 116, 327
Babylonians, 7
Baily's beads, 224, 231
Balloon studies, 156
Bartels, J., 326
Batteries, solar, 299
Becquerel, H., 259
Bell, B., 117, 283, 325, 326, 327, 328
Beryllium, 52
Betelgeuse, 274
Bethe, H., 263
Biermann, L., 127
Billings, B., 148, 150
Bipolarity of sunspots, 116–117